Dryland Opportunities

A new paradigm for people, ecosystems and development

About IUCN

IUCN, the International Union for Conservation of Nature, helps the world find pragmatic solutions to our most pressing environment and development challenges.

IUCN works on biodiversity, climate change, energy, human livelihoods and greening the world economy by supporting scientific research, managing field projects all over the world, and bringing governments, NGOs, the UN and companies together to develop policy, laws and best practice.

IUCN is the world's oldest and largest global environmental organization, with more than 1,000 government and NGO members and almost 11,000 volunteer experts in some 160 countries. IUCN's work is supported by over 1,000 staff in 60 offices and hundreds of partners in public, NGO and private sectors around the world.

www.iucn.org

About IIED

The International Institute for Environment and Development has been a world leader in the field of sustainable development since 1971. As an independent policy research organisation, IIED works with partners on five continents to tackle key global issues — climate change, urbanisation, the pressures on natural resources and the forces shaping global markets.

www.iied.org

About UNDP

UNDP Drylands Development Centre is part of the United Nations Development Programme. It is a unique global thematic centre that provides technical expertise, practical policy advice and programme support for poverty reduction and development in the drylands of the world.

The Centre's work bridges between global policy issues and on-the-ground activities, and helps governments to establish and institutionalize the link between grassroots development activities and pro-poor policy reform. The main areas of focus are mainstreaming of drylands issues into national development frameworks; land governance; marking markets work for the poor; decentralized governance of natural resources; and drought risk management.

www.undp.org/drylands

The Global Drylands Imperative

The Global Drylands Imperative (GDI) is a collaboration of organizations involved in drylands development. GDI includes the Canadian International Development Agency (CIDA), the United Nations Development Programme (UNDP), UNDP/GEF – Global Environment Facility, the International Institute for Environment and Development (IIED), the World Wide Fund for Nature (WWF), the International Union for Conservation of Nature (IUCN), and the Near East Foundation.

Dryland Opportunities

A new paradigm for people, ecosystems and development

Michael Mortimore

With contributions from:

Simon Anderson, Lorenzo Cotula, Jonathan Davies, Kristy Faccer, Ced Hesse, John Morton, Wilfrid Nyangena, Jamie Skinner, and Caterina Wolfangel

 This publication has been made possible in part by funding from the US Department of State Voluntary Contribution to IUCN.

Published by: IUCN, Gland, Switzerland, IIED, London, UK and UNDP, New York, USA

Citation: Mortimore, M. with contributions from S. Anderson, L. Cotula, J. Davies, K. Faccer, C. Hesse, J. Morton, W. Nyangena, J. Skinner, and C. Wolfangel (2009). *Dryland Opportunities: A new paradigm for people, ecosystems and development,* IUCN, Gland, Switzerland; IIED, London, UK and UNDP/DDC, Nairobi, Kenya. x + 86p.

ISBN: 978-2-8317-1183-6

Cover photo: IUCN Photo Library/Danièle Perrot-Maître

Layout: Gordon Arara (Publications Unit, Nairobi)

Printed by: NEUHAUS SA; Buenos Aires, Argentina

Available from: IUCN (International Union for Conservation of Nature)
Publications Services
Rue Mauverney 28
1196 Gland
Switzerland
Tel +41 22 999 0000
Fax +41 22 999 0020
books@iucn.org
www.iucn.org/publications

A catalogue of IUCN publications is also available.

CONTENTS

List of Figures

List of Tables

List of Boxes

Foreword

Drylands cover 41 percent of the earth's terrestrial surface. They are home to a third of all humanity, and have some of the highest levels of poverty, yet in most countries they have long been neglected by investment and sustainable development interventions. Drylands are disproportionately prevalent in poor countries, but they have been relatively marginalised from development processes and political discourse. This has allowed profound misunderstanding of drylands environments to become entrenched, leading to inappropriate and even detrimental interventions based on perceptions dominated by land degradation ('desertification').

The urgency of and international response to climate change have given a new place to drylands in terms both of their vulnerability to predicted climate change impacts and their potential contribution to climate change mitigation. There is a growing recognition also of the importance of dryland ecosystem services in supporting food security and other needs of dryland and non-dryland populations.

Externally driven, technical solutions for desertification and drylands development continue to be prescribed for problems that are highly complex and have social, political and economic dimensions. Such solutions may not only be unsuccessful in responding to the needs of dryland populations, but may, by disempowering rural dryland people, contribute to their marginalisation, thereby compounding the root cause of their poverty. A new paradigm is required that meets the needs of dryland people. It must address the full complexity and dynamics of dryland ecosystems, recognise their full potential for development, take account of changing world conditions, and restore the initiative to dryland peoples themselves.

This Challenge Paper builds on the understanding that has emerged over the past decade about climate dynamics in drylands and the role of uncertainty, risk and resilience. It situates this debate in the context of rapid global change - of climate, economy and geopolitics. The Challenge Paper emphasises adaptive potentials, the value of dryland ecosystem services and the investment and marketing opportunities they offer, and the possibilities of strengthening the institutional environment for managing risk and rewarding resilience. It aims to apply the new scientific insights on complex dryland systems to practical options for development. A new dryland paradigm is built on the resources and capacities of dryland peoples, on new and emergent economic opportunities, on inward investment, and on the best support that dryland science can offer. The authors recommend five building blocks: strengthening the knowledge base; valuing and sustaining dryland ecosystem services; promoting public and private investment in drylands; improving access to profitable markets; and prioritising rights, reform, risk and resilience.

This Challenge Paper presents a vision for drylands that makes their sustainable development a global rather than a local responsibility. The new interlocking of climatic and geopolitical factors means that drylands can no longer be treated as poor, remote, largely self-subsistent areas and left to their own devices. Through the recommendations presented in this paper, the Millennium Development Goals can be made more achievable, and biodiversity and ecosystem services can be maintained in the best interest of dryland peoples and the global community.

Julia Marton-Lefèvre	Camilla Toulmin	Philip Dobie
Director General	Director	Director
IUCN	IIED	UNDP/DDC

Acknowledgements

The authors are indebted to many people who, at different points in the evolution of this Challenge Paper, have offered critique or new perspectives, and especially to Professor B. L. Turner II, Professor Joseph Ariyo, Edmund Barrow, Dr Adama Faye, Dr Mike Norton Griffiths, Dr Eric Patrick, Anshuman Saikia, Dr Mark Stafford Smith, and Professor Jeremy Swift; to Dr Stefanie Herrmann for analysis of earth satellite data and creation of Figure 3; to Dr Wendy Strahm, our Editor; to Sarah Anyoti; Neville Ash; and to Gordon Arara for design and layout. The lead author is grateful to Caroline Edgar and Pat Hawes for their unfailing support through several consultancies for IUCN.

Acronyms

AVHRR............Advanced Very High Resolution Radiometer
CAMPFIRE.....Communal Areas Management Programme for Indigenous Resources
CARE...............Cooperative for Assistance and Relief Everywhere, Inc.
CBNRM...........Community-based Natural Resource Management
CDM................Clean Development Mechanism
CITES..............Convention on International Trade in Endangered Species of Wild Fauna and Flora
COP..................Conference of the Parties
DDC.................Drylands Development Centre (UNDP)
DDP.................Drylands Development Paradigm
DFID................Department for International Development
ECA..................Economic Commission for Africa
ECOWAS.........Economic Community of West African States
FAO..................Food and Agriculture Organisation of the United Nations
GCM................General Circulation Model
GDI..................Global Drylands Imperative
GDP.................Gross Domestic Product
GEF..................Global Environment Facility
GIMMS...........Global Inventory Monitoring and Modeling Studies
GTZ..................*Deutsche Gesellschaft für Technische Zusammenarbeit* (German Technical Cooperation)
ICIMOD..........International Centre for Integrated Mountain Development
ICRAF.............International Council for Research on Agroforestry
IIED.................International Institute for Environment and Development
ILRI.................International Livestock Research Institute
IISD.................International Institute for Sustainable Development
IPCC................Intergovernmental Panel on Climate Change
IUCN...............International Union for the Conservation of Nature
LGP..................Length of the growing period
MDG...............Millennium Development Goal
MEA................Millennium Ecosystem Assessment
NASA..............National Aeronautics and Space Administration
NDVI...............Normalised Difference Vegetation Index
NGO................Non-governmental organisation
NPP..................Net primary productivity
NTFP...............Non-timber forest product
PACD...............Plan of Action to Combat Desertification
PAGRNAT......Programme d'appui à la gestion des ressources naturelles de l'Aïr et du Ténéré
PES...................Payment for Environmental Services
SADC...............Southern Africa Development Community
UNCCD...........United Nations Convention to Combat Desertification
UNDP..............United Nations Development Programme
UNEP...............United Nations Environment Programme
USD.................United States Dollar
WRI.................World Resources Institute

CHAPTER 1
Drylands in a changing world

Since recession shocked the global economy in 2008, the meaning of 'sustainability' has taken on a new depth. Besides the long-running fear of environmental destruction – in particular in the 'susceptible drylands'[1] – and rising expectations of a 'tipping point' in climate change, a new question arises in the short run. Will 'business as usual' resume, with appropriate lessons learnt,[2] or will the system fracture? Can consumption continue to exceed sensible restraint, or can a new order emerge in the relations between humankind and the global ecosystem?[3] There are strong linkages between these questions. An additional question urgently needs answering: what will become of the Millennium Development Goals for the relief of poverty, injustice and inequity in the distribution of nature's benefits?

The scope of this Challenge Paper is the drylands, which include desert, grassland and savanna woodland biomes. One of the world's major ecosystems, the drylands have long lived with uncertainty and the threat of unsustainability, where moisture is scarce for all or part of the year, and soils for the most part infertile. This Challenge Paper is one of a series on the world's drylands.[4] It brings a perspective on conservation and sustainable development to particular approaches and strategies for development. It is argued that conservation – of biodiversity in particular – can only take place in healthy ecosystems, which in turn can only be maintained where poverty is reduced and appropriate institutions are operating.

Drylands matter

The following table shows that drylands occupy 41 percent of the earth's land surface and are home to 35 percent of its population.

The distribution of the world's drylands is shown in Figure 1. They occur in every continent, but are most extensive in Africa.

Table 1. The dryland system

Sub-type	Aridity index	Share of global area (percent)	Share global population (percent)	Percent rangeland	Percent cultivated	Percent other*
Hyper-arid	<0.05	6.6	1.7	97	0.6	3
Arid	0.05-0.20	10.6	4.1	87	7	6
Semi-arid	0.20-0.50	15.2	14.4	54	35	10
Dry subhumid	0.50-0.65	8.7	15.3	34	47	20
Total		41.3	35.5	65	25	10

*Includes urban
The aridity index is the ratio of precipitation to potential evapo-transpiration.[5]
Source: Safriel *et al.*, 2005.

[1] (UNEP, 1992)

[2] (World Bank, 2008)

[3] (Adams and Jeanrenaud, 2008)

[4] The previous Challenge Papers are eight in number and comprise an informal series issued by the UNDP Drylands Development Centre and other institutions associated with the Global Drylands Imperative over a number of years (Bonkoungou, 2001; Burton, 2001; de Oliveira *et al.*, 2003; Dobie, 2001; Dobie and Goumandakoye, 2005; Hazell, 2001; UNDP-DDC, 2001, UNDP-DDC, 2003)

[5] (UNEP, 1992)

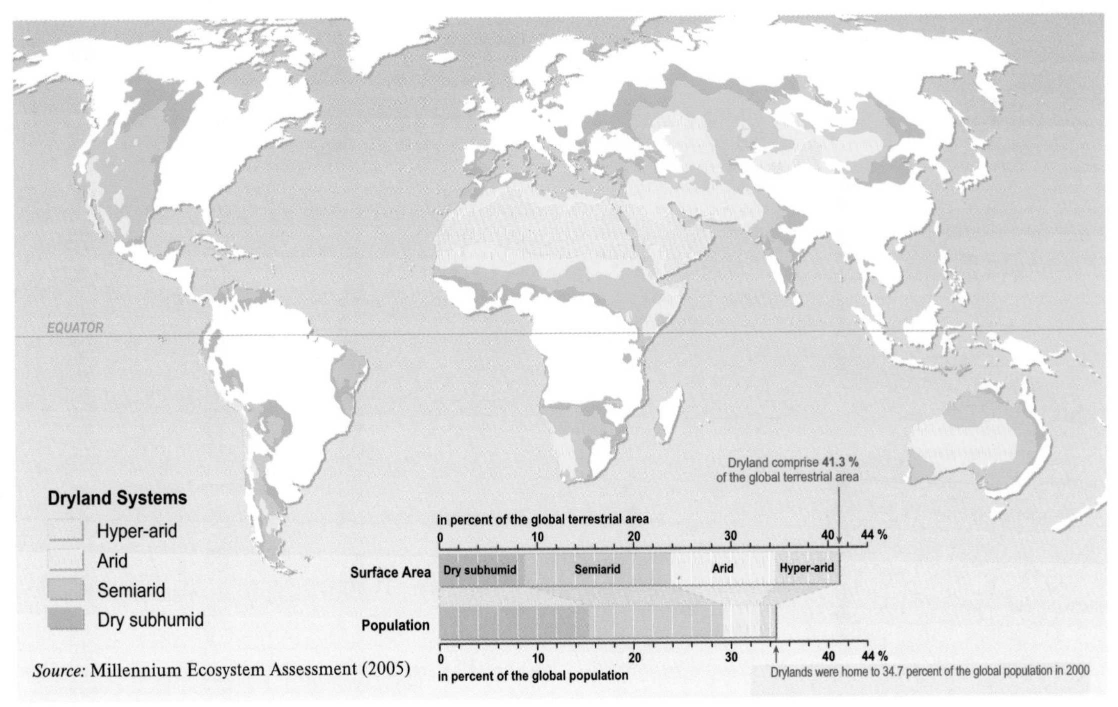

Figure 1: Distribution of the world's drylands according to aridity zones *(based on UNEP, 1992).*

Although the definition of drylands includes some arctic regions, these are not considered in this Challenge Paper, which confines its concern to tropical, subtropical and high mountain drylands. Their climates range from the hottest of tropical deserts to warm and cool temperate and high mountain regimes. In addition, drylands in some developed parts of the world (Australia, Europe, Israel, and the USA) are excluded. Despite sharing some significant commonalities (especially in Australia), many issues surrounding these drylands differ significantly from those of developing countries. Among these are the vast areas of deep rural poverty, particularly in India, China, much of Africa and Bolivia – one of the poorest countries in Latin America.[6]

The Millennium Development Goals provide global objectives for poverty reduction. Goals 1 and 7 have direct implications for the environment and therefore for drylands (Box 1). Faltering prog-

Box 1: Millennium Development Goals with direct implications for the environment

Goal 1: Eradicate extreme poverty and hunger.

Target 1: : Halve, between 1990 and 2015, the proportion of people whose income is less than $1 a day.

Target 2: Halve, between 1990 and 2015, the proportion of people who suffer from hunger.

Goal 7: Ensure environmental sustainability.

Target 1: Integrate the principles of sustainable development into country policies and programmes and reverse the loss of environmental resources.

Source: Dobie and Goumandakoye, 2005.

[6] There is no hard boundary between 'rich' and 'poor' or between 'developed' and 'developing' dryland countries. Relatively wealthy countries are excluded from this review in order to focus on poverty-environment linkages, mainly in the poor countries. Deserts (technically defined as 'hyper-arid') support about 100 million people, and although not excluded from this review, differ in significant ways from the more inhabited drylands. Their peculiarities are not fully discussed here.

Re-greening of the Sahel since the mid 1980's as a result of efforts of local farmers in the densely populated Maradi and Zinder regions in Niger. © *Chris Reij*

ress towards these and other goals – especially in dryland countries – increases the need for action in poor dryland countries. Such is the magnitude of dryland populations that failure in the drylands will mean failure for the global community. The benefits of conservation in dryland ecosystems - shown in this study – will make a major contribution to achieving the MDGs. Drylands must be central in strategies to achieve global sustainability. Six major challenges to global sustainability are identified in a recent study:[7] (1) poverty, inequity and human well-being; (2) globalization; (3) private-public balance in development; (4) environmental damage; (5) conflict and competition for resources; and (6) poor governance. All have their manifestations in the drylands.

The responses of dryland peoples to uncertainty, and their successes and failures in managing pressures on their ecosystems, are relevant in a world fearful of its future. They are likely to suffer disproportionately from the impacts of climate change engendered by others. However, due to their great extent, land use in the drylands can have an impact on atmospheric circulation and carbon fluxes. Among the world's major ecosystems, those of the drylands (in poor countries) have received less scientific and developmental attention in proportion to their size, their population, and their importance for global sustainability. They are 'investment deserts' in the struggle for wealth creation. They are inadequately understood by the world's policy makers and sometimes, even, by their own. In a few areas, severe and persistent conflict has been allowed to recur, with repercussions elsewhere.

[7] (Munasinghe, 2009)

Changing drylands, a changing world

The world as a whole has a stake in the health of dryland systems, not only because of their physical extent but on account of our increasing understanding of their interactions with global climatic, economic and geopolitical systems. Such forces are re-integrating drylands with global futures. Nowhere is this more obvious than in climate change, which forms a sub-text to all of the following issues.

Understanding dryland systems

Human-ecological systems are complex, co-evolve and interact.[8] The 'health' of the ecosystems is contingent on that of the human systems. Poverty is shockingly broad and deep in drylands, and ecosystem management is linked. Dryland ecosystems are considered to be under threat.[9] Land use change, which is at the heart of ecosystem change, is driven by policy, legislation, institutions and development interventions. A growing understanding of dryland dynamics only serves to underline their importance in the global system.

Land cover characteristics in drylands - because of their great extent - may influence atmospheric circulation systems well beyond the drylands. These characteristics result from millions of decisions made by users, a majority of whom are small-scale and resource-poor farmers, pastoralists and harvesters of natural products. Throughout Africa, data obtained from earth satellites show changes of unexpected direction and magnitude after 1980 (when the data series began).[10] These findings have precipitated a new debate about climate-society-ecosystem relations throughout the world's drylands.

Biodiversity is richer than sometimes thought in drylands, and both farmers and herders take an intense interest in natural diversity and agro-diversity, which takes on special significance during food shortages.[11] As protected areas are increasingly difficult to establish, maintain and police, accommodation must be sought between stakeholders and the health of their ecosystems.

Adaptive livelihoods

Many dryland peoples have developed resilience under hardship, variability, and risk that is based on historic and current adaptive knowledge and skills. Such skills are increasingly recognised, though it is commonly claimed that such capacities are not sufficient to cope with the speed of change, especially in the climate. Nevertheless, if better known and understood, they may contribute to development.

Urbanization, migration and population growth are in transition. Many drylands have doubled their resident populations in 30–40 years. Yet a demographic transition to lower fertility is slow to occur in many drylands. Urbanization is tipping the balance between urban and rural populations, and is rapidly approaching 50 percent in some countries. Dryland food producers may soon be outnumbered by urban consumers.[12] Ever more complex patterns of migration (local, regional, and international) are interlocking rural and urban economies, and many dryland households derive incomes from two or more places.

Under rapid urbanization, migrants take their human and financial capital with them to invest in housing, business and education.[13] This raises the opportunity costs of farm or livestock investments (e.g., in soil and water conservation). However, if they prosper, finance can flow in the opposite direction and benefit dryland ecosystems. Meanwhile, the supply of 'free land' is becoming exhausted. Farming depends on inheriting, buying, renting or otherwise appropriating land through markets and new institutional frameworks, both formal (legislated) and informal ('customary' and adaptive).[14] This leads to rising land values, subdivision or fragmentation of inherited land, and exclusion of the poorest. Grazing rights are threatened by expanding farms and weakening legal protection. Selling labour is driven by resource scarcity and the out-migration of rural labour. Asset portfolios in rural dryland households are increasingly entangled with other sectors and regions.

---

[8] (Reynolds *et al.*, 2007)

[9] (Adeel *et al.*, 2005; Safriel *et al.*, 2005)

[10] (Ecklundh and Olsson, 2003; Herrmann *et al.*, 2005; Olsson *et al.*, 2005)

[11] (Harris and Mohammed, 2003; Mortimore, 1989)

[12] For example in Senegal (Faye *et al.*, 2001)

[13] For example, in Kenya (Tiffen *et al.*, 1994)

[14] (Cotula *et al.*, 2006; Toulmin and Quan, 2000)

New opportunities for dryland farming

Policy makers and donors are beginning to prioritise agriculture after at least two decades of relative neglect. The World Bank promotes agriculture in its Annual Report for 2008.[15] Agricultural investments, after being neglected for two decades, are set to recover, though aid budgets may be affected by economic recession. But drylands have been subordinated to higher potential areas, a view that needs to be revised in the light of technological change and an increasing scarcity of land.

Improvements in plant breeding and other technological advances have potential to assist intensification in agriculture, although the gap between potential productivity and that achieved on farmers' fields remains large. Poverty, poor soils and high risk are major barriers. Scientific research is improving its capacity to take such barriers into account.

Raising soil fertility is often seen as the defining challenge facing dryland productivity. Major questions are not yet resolved concerning the options of an increased use of inorganic fertilizers, 'low external input agriculture', conservation ('no tillage') agriculture and soil biology strategies in drylands.[16] Healthy soil is a precondition of ecosystem health, including that of rangelands.

Biofuels are being promoted in some countries. Maize, soya and sugar cane (all grown in drylands, although sugar cane needs irrigation) are being produced on a large scale for manufacturing additives to petrol - for example, in Brazil.[17] Controversial questions about equity, profitability and technology surround this new 'opportunity'. Effects on food prices, loss of land (including rangeland) and other environmental impacts are poorly understood. Drylands are also attracting the interest of the Carbon markets, as although their sequestration potentials are low, their great extent makes them attractive. Financial transfers from polluting countries to the drylands would certainly have an impact on local economies. Other ecosystem services such as river basin conservation may also offer scope via payments for ecosystem services. But the danger is that markets will pre-empt a balanced policy analysis. Already corporate land grabbing for biofuel farming is taking place.[18] Certainly, these new technologies are unlikely to reduce hunger unless issues of entitlement (to resources, products and income) are confronted.

More integrated markets

Global economic recession in financial markets, manufacturing and trade has already weakened export-based development strategies, and may reduce migratory flows, while recent food price inflation gave cause for policy priority on national food security. Internal food commodity markets have been reinforced by urbanization. These developments suggest a need to overhaul natural resource-based strategies, and to make more sustainable and productive use of rural dryland ecosystems. Market liberalisation does not directly address the distinguishing feature of dryland markets, which is their limited capacity for stable output, a result of environmental variability. But grain staples and meat come from drylands, and investment should not be neglected by dryland governments.

Market communications (public or private investments in road and public transport, mobile phones and internet) are bringing markets more within the reach of dryland populations, including mobile pastoralists, with benefits to incomes, access to knowledge, welfare and quality of life. In the medium term, this trend will reduce the social or economic deprivation caused by living in drylands, especially if the provision of other services such as electricity, health, and education is speeded up (although not likely, perhaps, during conditions of economic recession).

Security and the environment

Geopolitical instability in some drylands cannot go unnoticed, and is alerting policy makers to linkages between security and ecology.[19] Increasingly integrated human-ecological systems bring such threats to the very doorstep of global capitalism (e.g., trans-Mediterranean illegal migration; Somali pirates; Afghan poppy production), and provoke military interventions and exhausting wars, in which dryland peoples are the primary victims. Responses at the national level are decreasingly effective as the issues become globalised. Security therefore must be factored in to governance models for the drylands.

[15] (World Bank, 2007)
[16] (Uphoff *et al.*, 2006)
[17] Luiz Inácio Lula da Silva, President of Brazil, in The Guardian newspaper (UK), 28 March 2009.
[18] (Cotula *et al.*, 2009)
[19] (Brown *et al.*, 2007)

A changing world needs changing drylands

Climatic interactions between drylands and global circulation systems (e.g., the export of Saharan dust to South America, the Caribbean, and even Europe; links between sea surface temperatures and African rainfall; el Niño effects on tropical rainfall) and geopolitical interconnections (e.g., effects of poverty on illegal migration to Europe; insecurity in ocean shipping lanes; international costs of dealing with food emergencies; terrorist incubation in mis-governed and impoverished dryland countries) are but a few reasons why the North cannot afford to ignore the drylands of developing countries.

This calls for a new paradigm, not only of dryland management (explored in Chapter 2), but also of international relations. But this should not lead to greater economic or technological dependency on the North. Experience has shown that rich countries' interests in drylands tend to be subject to short-term considerations - and may appear whimsical from a dryland perspective. Behind the arguments for a new and more equitable relationship with industrial-urban economies lies a stronger need than ever for dryland resilience in the face of variability.

Knowledge in dryland development

The dynamic elements outlined above compel a reappraisal of dryland futures. Their problems can no longer be represented as merely local or simply soluble with new technologies. A raised profile for the drylands on the global stage is an opportunity for better understanding and knowledge sharing among stakeholders (communities, governments, NGOs, donors, international conventions including the UNCCD, and others). While not ignoring past and present constraints, this Challenge Paper is dedicated to exploring such opportunities.[20]

Knowledge is a key component of the human system, and of the interactions between human and ecological systems that lie at the heart of dryland management.[21] But understanding the challenges of sustainable development has been impeded by a number of major misconceptions. It is the aim of this Challenge Paper to show that such perceptions, if applied indiscriminately, function more as myths than science. They are presented below, and we shall counter them in the following chapters.

1. **Dryland biomes - compared with other major biomes – are poor, remote and degraded, and apart from having tourist potential, do not really matter globally.** On the contrary, dryland issues are rapidly increasing in their global significance and call for international economic and institutional response [Chapters 1 and 8]. Table 1 shows the importance of dryland ecosystems to the world's population.
2. **Drylands are on the edges of deserts and the deserts are expanding ('desertification') owing to human misuse of the environment (overgrazing, deforestation and over-cultivation).** In place of this view of remorseless degradation, we propose a more balanced view of environmental management based on the concept of resilience [Chapter 2].
3. **Dryland peoples are helpless (their knowledge and adaptive capacity are weak) in the face of climate variability and change.** In place of despair, we situate drylands objectively within the climate change scenarios and argue that existing adaptive capacity, assisted by sound policy and research, can offer pathways to development [Chapter 3].
4. **Because of their low biological productivity (when compared to other major biomes), drylands have little economic value except to provide subsistence to those who live there.** We show instead, the real (or total) value of dryland ecosystem services both to local peoples' livelihoods and to national economies [Chapter 4].
5. **Drylands cannot yield a satisfactory return on investment owing to high risks resulting from low and variable rainfall.** We show that investment in drylands can and does yield a satisfactory and sustainable return, and that poor peoples' private investments are real and significant [Chapter 5].

---

[20] The literature on drylands is uneven, with an overwhelming dominance of African studies. This has resulted in a 'model' of dryland development that is strongly influenced by African experience. Information has been sought on drylands in Asia, South and Central America, but given their extent and diversity, the global drylands are difficult to generalise. Only scant attention has been given to literature in languages other than English or French, and to work published within Latin America and Asia. The reader is advised to keep this limitation in mind. A contributory study was commissioned on the status of drylands ecosystem services in Central and South America (Linares-Palomino, 2009).

[21] (Reynolds *et al.*, 2007)

6. **Drylands are weakly integrated into markets and because of their remoteness, poverty and low biological productivity, will remain so.** It can be shown that dryland communities have long used markets to drive development, that this economic strategy is expanding rapidly in importance, and that markets can function even under conditions of uncertainty [Chapter 6].

7. **Dryland communities are conservative and resistant to modernization and institutional change. Governance, rights and institutions are of only local importance and can safely be ignored in favour of new technologies.** We show that equitable rights (in particular, rights to the use of natural resources) and institutional change are necessary and achievable conditions for dryland development [Chapter 7].

8. **Risk and vulnerability resulting from uncertainty and environmental change can be adequately countered by standard development policy.** Instead, new approaches to risk management are emerging, which build on local and customary practice and directly confront variability [Chapter 7].

Figure 2 provides an illustrative model of the knowledge dynamics affecting dryland systems. At the centre is the biophysical dryland ecosystem, the source of ecosystem services (or natural resources), and the primary focus of its human resource managers. Scientists and policy makers have often tended to treat it as a quasi-closed system, and have tried to solve poverty reduction and conservation problems by technological interventions. The interaction with human ('socio-economic') systems, and in particular with the knowledge systems that inform and govern them,

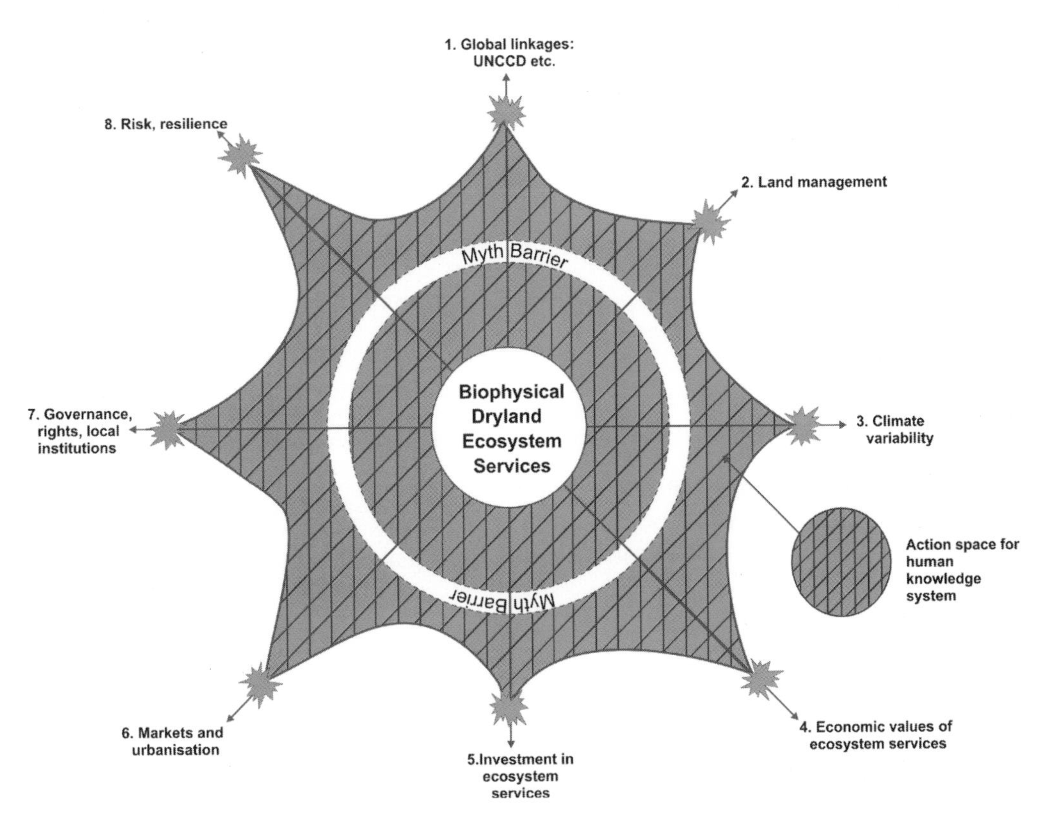

Figure 2. A model of the expansion of the 'action space' of dryland human-ecological systems.

A model to illustrate the expansion of a dryland human-environment system driven by knowledge. Each thematic trajectory represents 'progress' over time in expanding the 'action space'(room for manoeuvre, livelihood options, well-being) sustainably, based on the core ecosystem services. But a barrier – expressed in the 'myths of dryland development' – impedes progress. The point reached (shown by a flash) is a compromise between (a) local knowledge, capacity and opportunity, and (b) policies, institutions and interventions, which may still be influenced by residual 'myths'. Thus expansion is irregular and contradictory. The aim of dryland science and policy should be to remove impediments on all eight trajectories.

is shown as a shaded area of variable radius – an 'action space' (room for manoeuvre, livelihood options, well-being). Until recently, expansion of this action space was impeded by 'myths' shown as encircling it. These 'myths' frustrated progression of the dryland human-ecological system along defined trajectories (numbered 1 to 8). For example, the belief that dryland pastoralists and farmers were mainly land degraders prevented local knowledge and resilience from playing an effective role alongside science in following the sustainable land management trajectory (2). Removal of this impediment allows policy support for dryland systems to interact equitably with global markets, technology and financial resources. Human systems in drylands have engaged with the outer world historically – shown as 'progression' along the trajectories - and such engagement is now accelerating as the 'myths' weaken, but unequally. For example, progress is more rapid on the 'markets and urbanization' trajectory (6), where migration, income diversification, and education have spontaneously taken effect – unlike the 'investment' trajectory (5), where many dryland systems have been neglected by public investment owing to a persistent 'myth' that investment does not pay. The representation attempted here is only illustrative and subjective, but suggests that progress is variable. Over time the 'action space' occupied by the dryland human and knowledge systems will continue to increase.

Implicit in the idea of an expanding 'action space' in the present context is an interaction between the trajectories – for example, between climate variability (3), risk (8), and investment (5). Thus ecosystem services are used within a unitary whole, rather than within different sectors such as agriculture and forestry.

The aim of this illustrative model is to suggest how important knowledge systems are in shaping opportunities available to drylands, and to demonstrate the irreversibility of expansion as dryland human-ecological systems develop and interact with global systems. Old simplistic 'technology transfer' models do not capture the complexity of this process. This Challenge Paper attempts to confront the role of knowledge in defining the scope of the human systems, with specific reference to the eight trajectories (not an exclusive or final listing). Such an approach is strongly relevant to development policy, which also needs to escape from the barriers imposed by outdated, incomplete or misleading knowledge.

Key findings

What opportunities exist for dryland peoples and their ecosystems? What should a sustainable development framework look like? A strategy is needed that will achieve three aims: enhancing the economic and social well-being of dryland communities, enabling them to sustain their ecosystem services, and strengthening their adaptive capacity to manage environmental (including climate) change. In Chapter 8, an integrated strategy for dryland peoples and their ecosystems is proposed, based on the following major issues:

- Upgrading the knowledge base, improving knowledge sharing, and closing the gap between science and development practice in order to make best use of technology and to foster sustainable management. This includes improving understanding of dryland ecosystems (e.g., seasonality, variability, ecosystem services such as water, and human or social systems) [Chapter 2].
- Reassessing the total economic value of ecosystem services, to correct systemic undervaluation in national planning and policy, and improve well-being [Chapter 4].
- Promoting sustainable public investments in natural resources, to reverse decades of relative neglect, provide better incentives for private investment, and recognise the contribution of small-scale environmental investments [Chapter 5].
- Turning the growth of markets into an opportunity to remove barriers to participation, and to use more efficient, accessible and equitable markets as a pathway to sustainable development [Chapter 6].
- Supporting institutional changes to strengthen rights to natural resources, reform inequitable distribution, better manage risk, and increase resilience in the human-ecological system [Chapter 7].

Camel herder in Thar Desert. © *Jal Bhagirathi Foundation*

CHAPTER 2
A new paradigm of dryland development

Drylands are widely associated in the public mind with the complex and inadequately understood process known as 'desertification'. This association arises naturally enough from the proximity of many drylands to deserts, whose bounds tend to oscillate over time, periodically generating concern about 'desert advance'. In spite of such natural oscillations, human agency has taken the greater part of the blame. However, the degradation of some dryland ecosystems a long way from deserts calls for a broader concept and one that is inclusive of both climatic and human agency. Thus the United Nations Convention to Combat Desertification (UNCCD) defines desertification as:

'Land degradation in arid, semi-arid and sub-humid areas resulting from various factors, including climatic variations and human activities.' [22]

The degradation scenario, with the influential backing of the UNCCD, has dominated scientific understanding of dryland ecosystems. However its shortcomings have provoked a counter-paradigm that focuses instead on the resilience of dryland ecosystems under certain conditions. This unresolved scientific dialogue provides the frame of reference for this chapter.

The desertification paradigm

Theoretical basis

The implicit basis of the desertification paradigm is the idea of an ecosystem at equilibrium in which a perturbation is followed by a natural readjustment back to a stable state. Thus vegetation always evolves, through natural succession, towards its 'climatic climax'. However, human agency, through 'misuse' of the land (such as over-cultivation, overgrazing, deforestation, or excessive irrigation), may dislodge the ecosystem irreversibly from its former equilibrium. That is to say, its 'carrying capacity' - of livestock or of humans – is reduced. Restoration is not economically possible within an acceptable time-frame. The common use of the term 'fragile' denotes the susceptibility of dryland ecosystems to such degradation.

Evidence of reduction in ecosystem services – soil, forest, grasslands, water

The term itself has been used by different authors to refer to a range of changes in the state of ecosystems, such as rangeland degradation, deforestation in dry woodlands, soil nutrient depletion and erosion under farming, salinization under irrigation, a decline in biomass productivity (or net primary productivity, NPP) and hydrological desiccation, either on the surface or underground.

The biological productivity of an ecosystem depends on the health of its soils and soil moisture. Although the status of soils is commonly represented as unambiguously measurable in terms of key chemical nutrients, its productivity is also a function of physical and biological attributes (the latter still imperfectly understood). In Africa, a dominant narrative of soil degradation and erosion is especially influential in debates about dryland management. Early surveys claimed that 332 million ha (25.8 percent of the surface of Africa) are affected by soil degradation in the arid, semi-arid and dry sub-humid agro-ecological zones.[23] Estimates were published of the annual depletion of chemical nutrients which were upgraded and promoted by the World Bank and other agencies. These put net combined N, P and K losses at 60-100 kg/yr and increasing.[24] This narrative continues to guide policy makers, for example at the Abuja Fertilizer Summit,[25] notwithstanding its critics.[26]

[22] (UNCCD, 1993)

[23] (Oldeman and Hakkeling, 1990; Stoorvogel and Smaling, 1990)

[24] (Henao and Banaante, 1999; World Bank, 2003)

[25] (African Union, 2006)

[26] (Faerge and Magid, 2004; Mortimore, 1998; Mortimore and Harris, 2005; Scoones and Toulmin, 1998)

A large proportion of the world's tropical and subtropical forests are found in drylands. Dry forests are affected severely by deforestation in Africa (Sahel, Ethiopia and south-eastern Africa), and in parts of Latin America (Mexico, Peru, Paraguay), where annual rates of loss exceed 0.5 percent.[27] However, the removal of woodland may be compensated partly by regeneration of trees on farmland. A study in northern Burkina Faso found that while agricultural expansion is the main driver, such clearance is complex, nuanced and variable, discouraging generalisation.[28]

However, many drylands lack woodland. According to the FAO's statistics, most dryland countries in all continents (excluding deserts and with the exception of India and Pakistan) had from 30 to 50 percent of their area under pasture and fodder crops in 2000, and in many, this fraction had increased since 1980. In Central Asia, the fraction was higher. If fodder crops are excluded, in Africa there was a decline from 31.1 percent to 29.6 percent, owing mainly to agricultural clearances, whereas it increased in Asia (from 28.9 percent to 31.5 percent), America and Europe.[29] Such statistical changes, though small, represent large areas. Half of the Tibetan Plateau is considered by Chinese scientists to be degraded through overgrazing, and some areas are at risk from climate change.[30]

Structural or specific degradation in woodlands and grasslands or deterioration in the productivity of grasslands are alleged to be happening widely. The stability of grasslands is important (Box 2). With regard to 'scoring' biodiversity loss in drylands, however, data are reported to be scarce, and threats are perceived from changes in habitats, such as urbanization, which may overpower the 'legendary resilience' of dryland ecosystems.[31] But agrodiversity is valued by farmers (as shown in case studies of seed management),[32] and small-scale farming is far less destructive of biodiversity than large-scale mechanised systems.

By definition, water is scarce in drylands. Existing water shortages are projected to increase owing to population growth, land cover change and global climate change.[33] Water access and poverty are linked closely,[34] and when quality is taken into consideration, very large proportions of rural dryland people have restricted access. In Central Asia, and especially the Aral Sea basin, the restoration of rational water use planning is at the heart of development strategy.[35] According to a recent global analysis, there is a diversity of strategies for obtaining water, much variability in quality, and a variety of developmental challenges across 13 'water zones'; agriculture is the major user of water and more water harvesting and conservation farming (to reduce irrigation losses) are needed.[36] Land and water degradation are closely linked, so that improved land management can improve both livelihoods and water access simultaneously.[37]

The assessment of desertification processes has been beset (until recently) with problems of data quality and comprehensiveness. The first attempt at mapping soil degradation on a global scale was based only on expert opinions, and is now out of date (1991).[38] Estimates based on this work claimed that soil degradation affected 20 percent of drylands.[39] A contemporary study (also based on informal data) estimated that 70 percent

Box 2: Stability of grasslands

The African savannas are an interesting example. Grazing and burning have long been believed to determine the balance of trees and grasses. Under heavy grazing, the remaining grasses cannot support the fires necessary to inhibit the growth of unwanted shrubs, which expand their canopies and reduce the grazing for livestock. Thus an increase in biomass paradoxically degrades the land in economic terms. A synthesis of data from 854 sites in Africa has confirmed that at above 650 mm mean annual precipitation, fire and grazing animals are necessary to restrict the growth of a woody canopy, but below this level, the density of trees is controlled by rainfall, and diminishes with it.

Source: Sankaran *et al.*, 2005.

[27] (FAO, 2000)

[28] (Reenberg *et al.*, 1998)

[29] (FAO, 2008)

[30] (Wilkes, 2008)

[31] (Bonkougou, 2001)

[32] (Meles *et al.*, 2009)

[33] (Safriel *et al.*, 2005)

[34] (Thornton *et al.*, 2006; WRI *et al.*, 2007)

[35] (CAREC, 2003)

[36] (IFAD/FAO, 2008)

[37] (Bossio and Geheb, 2008)

[38] (Oldeman and Hakkeling, 1990)

[39] (UNEP, 1992)

Box 3: Institutionalising desertification

The Sahel Drought of 1968-74 led to the UN Conference on Desertification (UNCOD) in 1977. Its major output was a *Plan of Action to Combat Desertification* (PACD), which was to be implemented by the UN Environment Programme. Evidence of desertification was soon discovered in all continents (except Antarctica), and the 'susceptible drylands' were claimed to include 40 percent of the earth's land area and to contain 2 billion people.

The PACD was slow to have an impact, owing to a combination of under-funding, scientific scepticism and weak political commitment. However the Rio Earth Summit of 1992 spawned an effort to raise the profile of desertification again, in the form of the UN Convention to Combat Desertification (UNCCD). It was ratified eventually by 183 countries, with a permanent secretariat, funding agency (the Global Mechanism), and programme of activities centred on Conferences of the Parties (COPs) every two years.

It is important to understand that the concept of desertification has been strongly influenced by these institutional interests, its popularisation by non-governmental organisations and media, and the perceived funding needs of developing countries. Much effort has been invested in the measurement, assessment and mapping of indicators. This has been driven by the needs of donors' projects to 'monitor and assess desertification'. The science of desertification and sustainability has moved on but the influence of scientific research on Convention activities is widely considered to have been inadequate, and unnecessarily so.

Sources: UNCCD, 1993; UNEP, 1977.

of drylands are subject to some form of degradation.[40] A study commissioned by the *Millennium Ecosystem Assessment* suggested only 10 percent, but using a broad concept of biological productivity, the MEA concluded that 'there is *medium certainty*' that the true figure lies between 10 and 20 percent (in 2005).[41]

History is relevant, as the idea of desertification has taken on a life of its own. The term itself was coined by the forester Aubréville in West Africa in 1949,[42] but the belief that the forests of West Africa were degrading goes back to the beginning of the colonial era. Its popularity tended to oscillate with wet and dry spells in the rainfall. Thus it was rather neglected in the wet phase of the 1950s and 1960s. But the Sahel Drought of 1968-74 – the first major food emergency to receive global media exposure – changed everything (Box 3).

The myth of an 'advancing Sahara' lives on in the minds of certain agencies (Box 4), notwithstanding 30 years of contradicting research.

Drylands seen from space

Earth satellite data provide a fresh perspective on degradation in drylands. They offer a compatible basis for global estimates. The reflectance values in key parts of the spectrum can be used as proxy indicators of biological productivity. Applying this principle, using the *Normalised Difference Vegetation Index* (NDVI), or 'greenness' index, to the African Sahel produced surprising counter-evidence to the orthodox view of progressive degradation.[43] A strongly significant increase was observed throughout this agro-ecological zone between 1982 and 2006 (Figure 3).

This confirmed earlier findings on the oscillations of the desert edge,[44] and as those findings suggested, the trend was found to have a positive relationship with rainfall, which was then recovering from the drought cycle of the 1980s. However there were some localised exceptions to the general trend, and the strength of the association with rainfall was variable. Both observations suggest a role for another driver – perhaps management - either positive or negative. The data therefore need to be supported by studies of land use change, in context, on the ground.

Box 4: the myth of an advancing Sahara

Illustrating the persistence of this idea contrary to recent satellite-based evidence (Tucker et al., 1991; Herrmann et al., 2005; Olsson et al., 2005), the United Nations Environment Programme (UNEP) notes in its report Sudan Post-Conflict Environmental Assessment: "An estimated 50 to 200 km southward shift of the boundary between semi-desert and desert has occurred since rainfall and vegetation records were first held in the 1930s. This boundary is expected to continue to move southwards due to declining precipitation." (p. 9). It then adds: "The vulnerability to drought is exacerbated by the tendency to maximise livestock herd sizes rather than quality...." (p. 10). The main sources for the first of these views (on p. 62) are fieldwork in the 1930s (Stebbing, 1953), and later, the Sudan National Plan to Control Desertification (no date, no source), which calculates desert creep at 100 km in the last 40 years.

Sources: UNEP, 2007; Government of Sudan (supplied by J. Swift).

40 (Dregne and Chou, 1992)

41 (Safriel *et al.*, 2005: p. 637)

42 (Aubréville, 1949)

43 (Ecklundh and Olsson, 2003; Herrmann *et al.*, 2005; Olsson *et al.*, 2005)

44 (Tucker *et al.*, 1991)

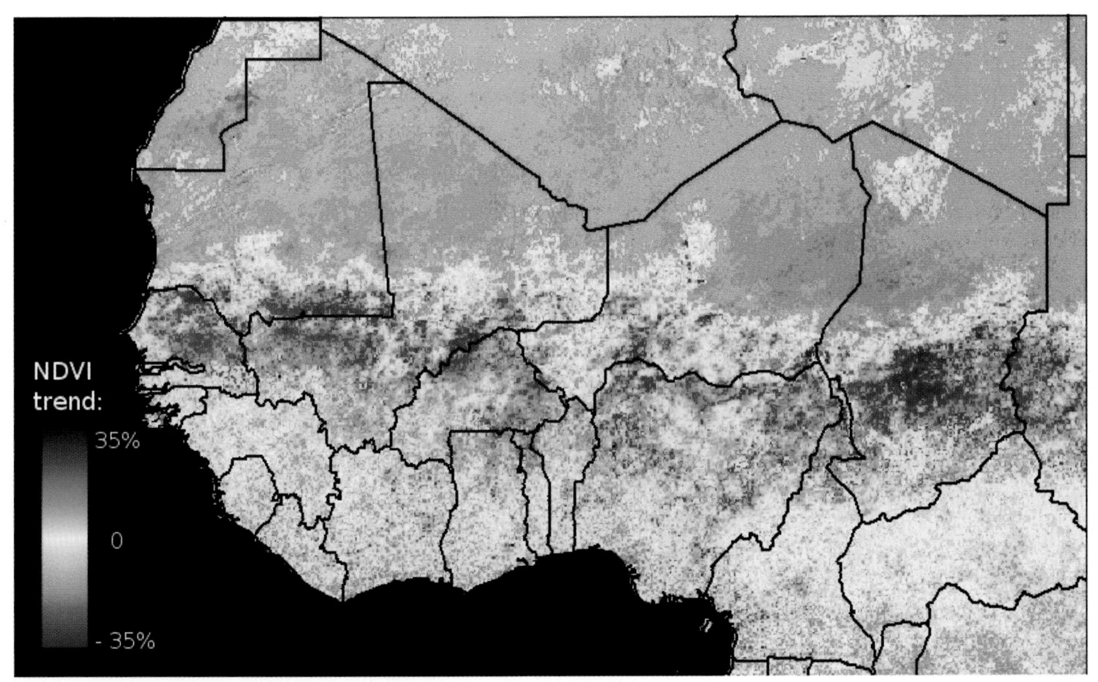

Figure 3. The 'greening' of the Sahel, 1982-2006.

Technical Note: Linear trends in the vegetation greenness index (NDVI) are shown in percentages. Trends were computed from monthly 8 km resolution AVHRR NDVI time series produced by the GIMMS group, NASA Goddard Space Flight Center, USA. *Source:* Extended from work previously reported in Herrmann *et al.*, 2005.

Other studies and data for other regions tend to strengthen the evidence of a relationship between vegetation 'greenness' and rainfall, leaving less space for the management drivers so often blamed for dryland degradation. A global synthesis of data on rapid land use change failed to confirm that the African Sahel was a 'hotspot' of desertification, and concluded that Asia has the greatest concentration of dryland degradation.[45] A study covering China-Mongolia, the Mediterranean, the Sahel, Southern Africa and South America found that 'a strong general relationship between NDVI and rainfall over time is demonstrated for considerable parts of the drylands. . .a 'greening up' seems to be evident over large regions'.[46]

Using NDVI data to estimate net primary productivity (NPP), the approach has been applied at a global scale, allowing conclusions to be drawn about changes in biological productivity throughout the world's drylands. This study found that during the period 1982-2006, global drylands (41.3 percent of the earth's land surface) contributed only 22 percent of the world's degrading areas (Box 4).[47] In fact, drylands do not figure strongly in ongoing land degradation, except in Australia. In Africa, the recovery of the Sahel from the droughts of the 1980s is a notable feature. Globally, there is little correlation between land degradation and the aridity index; 78 percent of degrading areas are in humid regions, 8 percent in the dry sub-humid, 9 percent in the semi-arid, and 5 percent in arid and hyper-arid regions.

These findings appear to question the common perception of the drylands as the major focus of land degradation. But the authors suggest that degradation is cumulative, and therefore, areas degraded before 1981 may have stabilised at low levels of productivity, while the data show *additional* degradation since 1981. This hypothesis, however, requires testing.

The drylands are better served by the NDVI (as an indicator of productivity) than the forests, where canopy formation is complex. However, the derived values for NPP still provide an imperfect proxy for the use-value of dryland farms or pastures. Some invasive plant communities, such as *Prosopis juliflora* and *P. chilensis*, are useless for grazing, or (like the indigenous species *Calatropis procera*) may indicate the abandonment of economic

[45] (Lepers *et al.*, 2005)

[46] (Helldén and Tottrup, 2009)

[47] (Bai *et al.*, 2008)

production on farmland. Standing stocks of timber which grow incrementally each year may be underestimated. Cycles of fallow and cultivation may give misleading signals of soil productivity. Crop production on farmland concentrates a major fraction of NPP within a short growing season of 3-5 months. However, total annual production of biomass (crops, fodder, fuel, compost) may compare favourably with that of natural vegetation under the same average rainfall.[48]

Earth satellite data - whose analysis is still at an early stage - draw attention once again to the dominant influence of variable rainfall on biological productivity, and call into question a simplistic notion of dryland degradation emphasising only 'human mismanagement'. In order to progress further in understanding evolving dryland ecosystems, a different model and change of scale (from global to local) is called for. This is discussed in the next section.

The resilience paradigm

Theoretical basis

The counter-paradigm begins with recognising that dryland ecosystems are not characteristically at equilibrium. As their productivity depends primarily on variable rainfall, they are better understood as in a state of disequilibrium.[49] For example, plant biomass in rangelands is driven by annual rainfall rather than by stocking pressure - for when pasture fails, the animals die or migrate. However seed banks in the soil ensure that vegetation recovers, though not necessarily with the same species composition. On some Sahelian rangelands, such as those of the Manga Grasslands on the Niger-Nigeria border, the dominant perennial grasses were replaced by annuals following the Sahel Drought of 1969-74.[50] This capacity of the ecosystem to maintain its functional integrity while adjusting to variable drivers justifies describing it in ecological terms as *unstable* and *resilient*.[51]

The same principle may apply more widely than in the drylands, and to social or economic systems as well as ecosystems. Evidence from interviewing African farmers in high-risk, drought-prone agro-ecosystems suggests that such a view corresponds more closely with the strategies employed to manage their livelihoods under conditions of uncertainty (Box 5).[52] In particular, the persistence of Sahelian farming livelihoods through three decades of declining rainfall (from the mid-1960s to the mid-1990s) provides support for such an approach.

A proposed *Drylands Development Paradigm* (DDP) offers a significant scientific advance from the worn conventional wisdom of 'combating' dryland degradation.[53] Its authors argue that recent advances in dryland development, together with integrative approaches to global change and sustainability, suggest that concerns about degradation, poverty, biodiversity and other issues can be confronted

Box 5: Degradation reversed in Machakos District, Kenya

Fundamental changes observed in Machakos District, over a period of 60 years, suggested positive linkages between population growth, market development and sustainable environmental management. In the 1930s and 1940s, government officials were extremely concerned about erosion on the hillside farms and clearance of dry woodland. Yet, the district saw the value of output per square kilometre increase sevenfold between the 1930s and the 1980s. On a per capita basis, a doubling in output occurred, even as the population increased fivefold.

The local farmers achieved this through a fundamental transformation in farming practices, including: a reversal of erosion thanks to the construction of thousands of kilometres of farm terraces and field drains; improved productivity through integrated crop-livestock production systems; new or adapted farm technologies; increased labour inputs; and increased private investments, which were financed in part from off-farm incomes. Established systems of land tenure, better dissemination of knowledge through women's groups and other flexible institutions, improved technology, access to urban markets and the relaxing of farm legislation all contributed to dramatic socio-economic improvements in the district.

Source: (Tiffen *et al.*, 1994)

---

48 (Mortimore *et al.*, 1999)

49 (Benkhe *et al.*, 1993)

50 (Mortimore, 1989)

51 (Holling, 1973; Holling, 2001)

52 (Mortimore, 1998; Scoones, 1994)

53 (Reynolds *et al.*, 2007)

with a renewed optimism. Among the principles emphasised are the co-evolving and co-adapting nature of human-ecological systems, the distinction between 'slow' and 'fast' variables of change, the importance of thresholds and scale and of local environmental knowledge. Consistent with the idea of resilience, it builds on case studies of endogenous management and adaptation rather than relying only on an exogenous technical viewpoint.

Scale and time

While a global perspective has tended to drive policy debates around dryland degradation (at least for the UNCCD and donors), any action on the ground must engage with local perceptions and livelihood priorities if it is to have any chance of success. Top-down interventions have often failed in the past. The effective empowerment of local communities to manage their ecosystems sustainably is a recognised aim in the successes that have been claimed.[54] It is at this scale that the disequilibrium of dryland systems can be best understood, and the adaptive capacities of communities to live with uncertainty.[55]

It is also necessary to develop a more sophisticated understanding of temporal change. The DDP proposes a distinction between 'slow' and 'fast' variables. 'Desertification indicators' measure, or try to measure, the state of a system at a point in time. Points, however, are ephemeral as slow system variables change, driven by environmental, economic and socio-political forces. More important, if fast variables (such as rainfall) are not at equilibrium, comparing 'snapshots' is logically meaningless unless the intervals can be mapped. Therefore, a painstaking analysis of system change - in the medium to long term - is necessary to expose both variability and trends, having positive lessons to teach and whose direction may offer opportunities for enabling interventions.[56]

A search for bio-physical, social and economic indicators faces the difficulty of finding a satisfactory datum against which the regression of a complex human-ecological system may be measured. When we look at the cumulative history of human use of the earth's surface (stretching back several millennia in regions such as China, India and the Mediterranean), and the very recent extension of new management regimes in almost all drylands, inhabited or not, it becomes impracticable to base present policy objectives on restoration of a landscape of pristine ecosystems.

Evidence of resilience in dryland ecosystems

The desertification paradigm cannot account for the long term persistence of pastoral and smallholder farming systems in drylands which largely support populations many of which have doubled in about 30 years. They specialise in livestock, or use livestock in integrated crop-livestock systems, intensifying agriculture mainly through additional labour inputs, skills, organic fertilization, and increasing participation in markets.[57] A synthesis of pastoral rationale is summarised in Box 6. This rationale still drives pastoral systems despite many misguided attempts to 'modernise' them, deprive them of resource access, or marginalise them politically. Understanding, consolidation, dissemination, flexible and supportive policies are necessary if the best use of this knowledge is to be made both in development and in adapting to climatic variability.

Besides the well-documented example of Machakos in Kenya (Box 5), evolving intensification has been documented in Senegal, Burkina Faso, and northern Nigeria, where its roots go back several centuries.[58] Moreover, this process of incremental intensification is spreading rapidly in response to growing scarcities of land, for example in Makueni which is a more recent extension of the Machakos model in Kenya, and in extensive areas of Senegal and Nigeria. Even at the national scale, long term data (1960-2000) do not support theories of agricultural collapse. Rather, the intricate interactions of policy with production and yield from year to year suggest that the role of demand factors has been underestimated.[59] These interactions are difficult to unravel because the proximate determinant of yield in any year is the rainfall.

There is no doubt that nutrient levels decline on repeatedly cultivated soils in drylands, and fallows often do not fully compensate. But chemical fertilizers

[54] (UNCCD, 2006)

[55] (Safriel *et al.*, 2005: pp. 645-6)

[56] Grassland ecosystems used as rangelands represent a narrower focus where 'desertification' has a stronger technical basis and both ecosystem change and its management are better understood. These are sub-systems within drylands as a whole.

[57] (Mortimore, 1998)

[58] (Mortimore, 2005)

[59] (Djurfeldt *et al.*, 2005; Mortimore, 2003)

Seasonal changes – Shagarab, Gedaref State in semi-arid to arid North-Eastern Sudan. Sorghum field after the rainy season in October and the same area after the harvest in March. © *Caterina Wolfangel / Edmund Barrow*

Box 6: Resilience of pastoral systems in Africa

Risk is spread in the following ways, which enhances the resilience of the system:

The range: Livestock mobility, over space and time, optimizes use of the range where rainfall is spatially and temporally very varied. Large and diverse ranges comprising wet, dry and drought time grazing areas are managed as common property resources. Knowledge of when wild species, particularly trees, yield food helps to supplement reduced milk yields during dry times. Tree conservation is vital for conserving fodder, providing shade, and other benefits. Many (usually tree-based) products can be sold, for example gums, resins and medicinal plants.

Water: Water management is tightly controlled, and rights are negotiated, along with range management, and the availability of water often gives livestock access to valuable pastures.

Diversification: A diversity of animals (grazers and browsers) reduces risk from disease, droughts and parasites. Risk is further controlled by redistributing assets through mutual support, including splitting herds between pastures. Mitigating risk from drought may involve diversification into distant labour or trading markets, as well as expanding trade in wild products. Opportunistic rain-fed agriculture is practised to spread risk (the Turkana of Kenya have 23 sorghum varieties that only need 60 – 90 days to mature).

Institutions: Risk management, through diverse traditional institutions such as *Qaaran* in Somali, *Iribu* in Afar, and *Buusa Gonofa* in Borana, include ways to support those households that have lost livestock from drought, raids, and disease. These social safety nets enhance labour sharing and security during periods of stress.

Source: Barrow, 1996.

are only a part of the answer, and it is recognised that affordable and integrated fertilizer management requires amendments to be made in 'micro-doses' together with recycling organic matter. In intensive dryland systems, these are already common practices. Indeed, the term 'micro' goes to the heart of African and possibly all small-scale farming in drylands – it refers to the practice of treating plants individually with attention to the micro-variation within the field in soil and water conditions.[60] Also, attention is given to maintaining the biological properties of the soils. In extensive semi-arid farming systems, organic matter is transferred from rangeland to farmland by grazing animals. The amount of rangeland available, for example in the crop-livestock system of Niger, therefore appears to determine the numbers of animals and the supply of organic matter.[61] Paradoxically, however, in the much more intensive system of rural Kano, crop residues can support higher stocking densities even without rangeland.[62] Kano benefits from higher average rainfall (650 mm compared with 400 mm).

On the rangelands, an outsiders' polemic of degradation, assumed to be driven by overstocking, has been based on the concept of equilibrium – a theoretical carrying capacity. It argues for stocking levels to be controlled at the maximum supportable in the driest years. This ignores the key strategy of herd mobility. But from Africa to Central Asia, seasonal transhumance and year-round mobility have provided herders with a defence against disequilibrium, uncertain rainfall and pasture productivity. Opportunistic stocking strategies with associated risk-bearing make good economic sense in such an environment.[63]

Governments have been very slow to recognise the rationale of pastoral mobility and its implication – that over-grazing is caused by impeding movements with barriers, boundaries and regulation, rather than by allowing moving herds to redistribute grazing pressure and thereby better conserve the ecosystems.[64] In Niger and northern Nigeria, for example, early studies of WoDaaBe Fulani groups showed how social or kinship relations were reflected in their distribution in space during the rainy season, and seasonal variations in pasture condition determined their transhumance.[65]

In Mongolia under socialism, fixed territories and shared ownership of livestock were imposed on a cultural landscape that included four discrete resources: seasonal pastures, reserve pastures, hayland and sacred lands. Because collective management of the range was practised before, pastoral groups achieved a measure of adaptation. But when private ownership was introduced in the 1990s, together with open access to profitable markets for cashmere wool, new entrants were attracted to livestock herding who avoided transhumance in favour of clustering around fixed water points and settlements. Range degradation is now reported (Box 7). Such examples as these suggest that neither imposed planning controls nor unfettered market forces can substitute adequately for indigenous knowledge and practice; moreover the resilience of both an ecosystem and a long-practised mode of management may be put at risk by development interventions.

[60] (Brouwer, 2008; Mando *et al.*, 2006)

[61] (Schlecht *et al.*, 1998; Turner, 1999a)

[62] (Harris and Yusuf, 2001)

[63] (Sandford, 1994)

[64] (WISP, 2007a)

[65] (Dupire, 1962; Stenning, 1959)

Box 7: Institutions, markets and environmental change in Mongolia

In Mongolia, traditional pastoral communities managed their cultural landscapes by means of mobility through the four seasons of the year. By regulating access to rangeland, these systems were sustainable for centuries. During the socialist era (1950–1990), rangeland ownership was vested in the state, and the pastoralists were organised into collectives. Use of traditional cultural landscapes survived generally during the socialist period, although there were changes in movement frequency and distance. With the transition from socialist collectives to a market economy, there was a revival of traditional pastoral networks (at *hot ail* level) which regulate labour and access to grazing. However, while the rangeland still belongs to the state, the privatisation of animal ownership together with rising prices for cashmere attracted many new households to set themselves up. The goat population increased from 5.1 million in 1990 to 18.3 million in 2007 in response, and the number of herders more than doubled. Many of these households were inexperienced and insufficiently mobile; they settled near water sources or settlements where the carrying capacity of the grazing resources is now considered to be exceeded. More than half have less than 100 animals and are at risk from poverty. The landscape was fragmented, depriving many communities of one or more of their seasonal ranges. A period of grazing scarcity owing to heavy snows (*zud*) in 1999-2002, which caused heavy livestock losses, increased migration from outlying areas to the vicinity of the cities. However, warmer temperatures (by 1.94°C during the last 60 years) and reduced precipitation in spring are adversely affecting water and forage resources.

Sources: (Chuluun, 2008; Ojima and Chuluun, 2008).

Deforestation in dry forests – or 'savannization' in West Africa – is linked in the minds of foresters with burning, often called 'indiscriminate'. Forest species are considered to be at risk of substitution by fire-resistant species in an open savanna. Burning takes place in the dry season, and in southern Mali up to 57 percent of the landscape may be burned each year.[66] But analysis has shown that burning is practised on a micro-scale - a 'patch-mosaic' –and carefully managed to conserve ecosystem productivity.

Agricultural clearances of dry forests have often been replaced (in Africa and India) by the culture of planted or self-generating trees of economic value on private farmlands. Even as some officials were predicting a treeless desert around major population centres, and enforcing draconian forestry law against woodcutting, an increasing scarcity of both timber and non-timber forest products was valorizing the trees in the livelihoods of many rural communities. Thus the well-timbered farmlands surrounding Kano in Nigeria retained or increased their stocks through the two drought cycles of the 1970s and 1980s, when selling fuelwood was a major temptation for food-deficit households.[67] More recently, eastern Niger has witnessed a dramatic regeneration of indigenous trees on farms.[68] Improved rainfall has helped, but the story illustrates resilience in the human-ecological system, facilitated by sound policy.

Conclusion: reversing the priorities

Dryland development is thus not only about ecosystems. Dryland peoples score well behind others in indicators of poverty, vulnerability, and wellbeing.[69] This tends to be true at global, regional and national levels. For example, dryland areas in Argentina and Brazil have twice the national average percentage of poor and indigent people; and two-thirds of poor or indigent Brazilians live in the north-eastern region, which is mainly dryland, half of them in rural areas.[70]

Are human poverty and environmental degradation different problems or one and the same? The debate on the causes, rate and extent of degradation in dryland ecosystems is by no means closed. However the necessity for a 'people-centred' model is widely conceded. Hitherto development was understood to be conditional on achieving environmental sustainability. Thus technical approaches dominated donor and government concerns. But the resilience paradigm reverses this dominance in favour of development as a condition for sustainability. Thus the misguided idea of a fundamental trade-off between development and sustainability has been exposed by the achievements of dryland people themselves. Their 'success stories' are replicable, given an enabling policy environment. Development emerges as the key goal, not only in poverty and risk reduction, but also in reversing degradation. In Chapters 4 - 7, this vision will be explored in terms of the following themes: valorizing dryland ecosystems (4), restoring investment (5), linking up with effective markets (6), and rebuilding institutions (7). In doing so, an ideology of simply 'combating' degradation is rejected as a sound basis for policy. Before doing so, however, the critical issue of climate risk and change must be confronted.

---

[66] (Laris and Wardell, 2006)

[67] (Cline-Cole *et al.*, 1990; Cline-Cole, 1997)

[68] (WRI, 2008), 142 *ff.*

[69] (Dobie and Goumandakoye, 2005)

[70] (Linares-Palomino, 2009)

A herdsman in El Beyyed, Mauritania, in front of his private stone-age artefact museum. This herdsman is volunteering his time to patrol the area to protect it against thieves. © *Piet Wit*

CHAPTER 3
Adapting to climate risk and change

Adapting to climate risk in the past

Change and variability are intrinsic properties of most dryland climates. History and archaeology record evidence of phases of desiccation, when deserts expanded into surrounding semi-arid grasslands; and phases of desert retreat, when human activities such as hunting, grazing and fishing spread into what are now hyper-arid environments. Between such phases there were periods of relative stability, but rainfall still varied between years. A recent example was the southward extension of the Sahara Desert, associated with the great drought cycles of the early 1970s and 1980s, which was followed by a northward advance of the vegetation after 1980.[71] Evidence from Lake Bosumtwi, Ghana (including geomorphic, isotopic and geochemical data), shows that the severe drought of recent decades was not anomalous in the past three millennia, and that the West African monsoon is capable of longer and more severe droughts still.[72] Drying trends – which have not yet reversed – have been observed in other large regions between 1900 and 2005: in the Mediterranean, southern Africa, northwest Mexico and northwest India.[73]

Uncertainty exists, therefore, as to whether recent drought cycles are attributable to recent anthropogenic global warming.[74] Take for example, the West African Sahel, where rainfall records indicate periods of wetter and drier conditions, each lasting for several decades spread over half a century (1941-2001), and interspersed with periodic harsh droughts (Figure 4). It is not known whether such a pattern is influenced by global warming, but relationships with the El Niño Southern Oscillation and with changes in sea

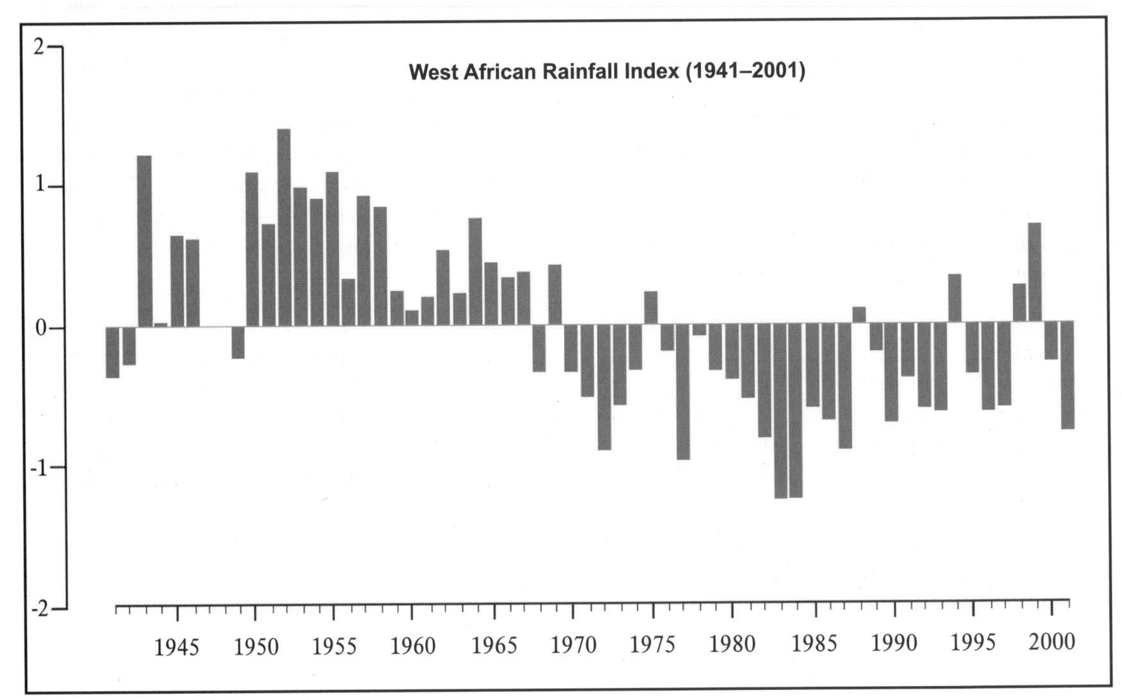

Figure 4. Variability in West African rainfall, 1941-2001. Source: DFID (2006).

[71] (Tucker *et al.*, 1991)

[72] (Shanahan *et al.*, 2009)

[73] (Anderson *et al.*, 2009)

[74] (Trenberth *et al.*, 2007 p. 255-6; IPCC, 2007)

surface temperatures have been observed. Possible influences of land degradation or biomass burning have yet to be demonstrated.

Southern and Western Africa have seen an increase in the number of warm spells and a decrease in the number of extremely cold days. In East Africa temperatures have fallen, close to the coasts and major inland lakes.[75] Available evidence suggests that Africa is warming faster than the global average and is likely to continue to do so.

Sub-Saharan African farmers' adaptive strategies - whether forced on them by food insecurity or adopted opportunistically - are well documented.[76] They include:

- ***Agricultural:*** shifting between crops, varieties, specializations, in response to rainfall, market or other changes. For example, in the Sahel warming plus reduced rainfall has reduced the length of the vegetative period 'no longer allowing present [long cycle] varieties [of millet] to complete their cycle'.[77] Similar challenges are reported for other crops.[78] However, Sahelian farmers usually cultivate both long and short cycle millets with the aim of spreading risk. This means that they have been able to adapt their cropping patterns to shifts in rainfall over recent decades.[79] In northern Ethiopia, farmers have shifted to more drought-resistant crop varieties to shorten the cropping calendar and accommodate less rain in the spring and summer (even though rainfall records show no downward trend).[80]
- ***Livestock:*** changing seasonal grazing migrations to take advantage of alternative forage when their usual grazing is damaged by drought. For example, during the droughts of the early 1980s, cattle herds of the WoDaaBe Fulani migrated from Niger into Nigeria, where they had not been before, consuming forage resources of the semi-sedentary Fulani who did not resist them.[81] Herders' adaptive strategies in Eastern Africa have included accessing tree fodders and powered boreholes, selling animals, and intensifying animal health care.[82]
- ***Social claims:*** the destitute claimed assistance from community leaders, traders or kinsmen, or as a last resort, took to begging. But social obligations were already eroding in the Sahel in the 1970s.[83]
- ***Seeking wild foods:*** women in particular are repositories of local knowledge on the use of a range of ecosystem products which, when times were hard, might replace food grains, though at a considerable cost in time and health.[84]
- ***Income diversification and migration:*** farmers took to harvesting natural products, adding value through manufacturing simple objects, labouring and otherwise exploiting the multiple - though poorly rewarded - work opportunities offered locally, as well as migrating further afield. However, in northern Ethiopia, famous for its drought-linked food shortages, vulnerability does not necessarily make a person migrate as other options are explored first.[85] More opportunistically, a severe food shortage in the 1970s sent young men from the northern borderlands of Nigeria 1200 km to Lagos to seek temporary, low paid employment.[86] This later developed into a profitable regional trade in livestock.

Such strategies are not equally open to all, and opportunity to use them depends not only on knowledge, but on resources in the local community. Adaptive capacity in Senegal was found to be undermined by poor health, rural unemployment, and inadequate village infrastructure.[87] How adequate is this resource to face new challenges, whether climatic or economic?

---

[75] (Boko *et al.*, 2007)

[76] (Rahmato, 1991)

[77] (Ben Mohammed *et al.*, 2002)

[78] (Van Duivenbooden *et al.*, 2002; Rosenzweig *et al.*, 2007)

[79] (Brock and Ngolo, 1999; Roncoli *et al.*, 2001)

[80] (Meze-Hausken, 2004)

[81] (Mortimore, 1989)

[82] (Morton, 2006)

[83] (*Ibid.*)

[84] (*Ibid.*; Harris and Mohammed, 2003)

[85] (Meze-Hausken, 2000)

[86] (Mortimore, 1989)

[87] (Tschakert, 2007)

June–July–August (JJA)

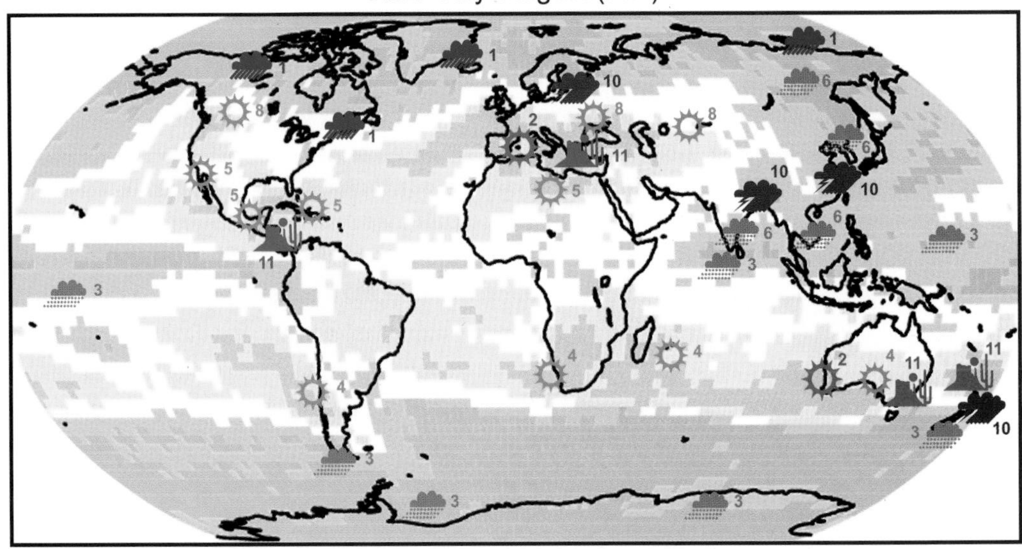

December–January–February (DJF)

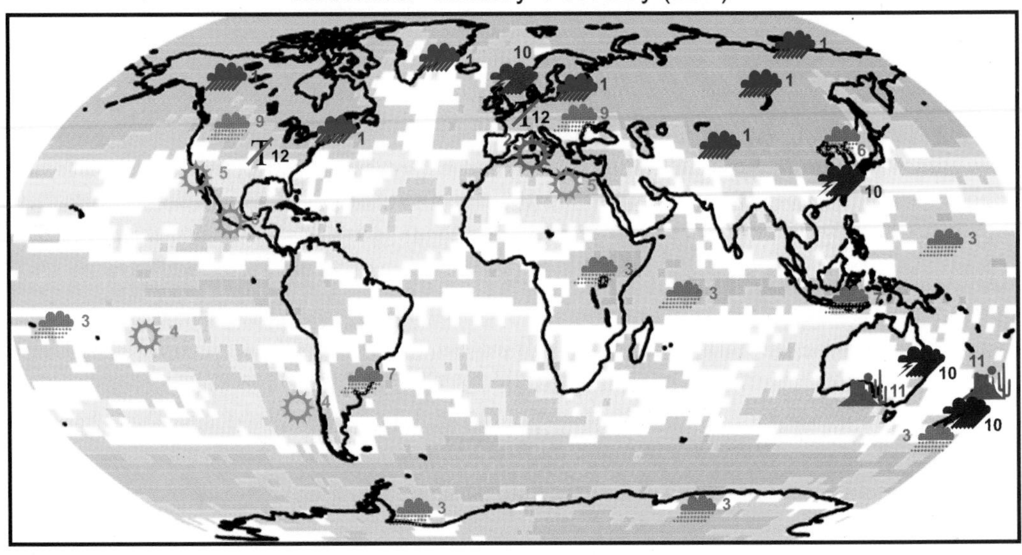

Figure 5. IPCC projections for mean change in precipitation from the periods 1980-99 to 2080-99.

Note: The two panels correspond to June-July-August (upper) and December-January-February (lower). The colour shading shows the fraction of the 21 GCM simulations that predict precipitation increase (blue, ≥90%, or green, ≥66%) and decrease (yellow, ≥66%. or brown, ≥90%). The maps may be compared with the global distribution of drylands (see Figure 1). *Source:* IPCC Working Group I[88].

[88] (Solomon *et al.*, 2007, Chapter 11, see p. 859 for a full explanation of the maps including the symbols shown).

Adapting to climate change in the future

Although some drylands will be significantly affected by temperature changes, hydrological changes will have the greater impact (with the exception of dry mountain climates). Predictions for different regions vary considerably among 21 General Circulation Models (GCMs) used by the Intergovernmental Panel on Climate Change (IPCC) in its Fourth Assessment. A guide to the probability of a predicted increase or decrease in average precipitation is the fraction of these models that agree. Figures 5 and 6 show these fractions, for December-January-February and June-July-August. The major drylands of the world can be located within these patterns (compare Figure 1).

The maps show that opposite trends are expected in different regions and that the levels of confidence we can place in them also vary. The diversity of outcomes expected under climate change is illustrated by the following trends identified by the IPCC in some regions which include major drylands (months are shown in parentheses):[89]

- Very likely winter (DJF) increase in the Tibetan Plateau
- Very likely annual mean decrease in most of the Mediterranean area
- Likely annual mean increase in tropical and East Africa
- Likely winter (JJA) decrease for southern Africa
- Likely annual mean decrease in North Africa, northern Sahara, Central America
- Likely summer (JJA) increase in East Asia, South Asia and most of Southeast Asia
- Likely winter (DJF) increase in East Asia
- Likely summer (JJA) decrease in Central Asia
- Likely increase in the risk of drought in the Mediterranean and Central America

It will be clear that the impacts of climate change in future will be highly specific to region and cannot be generalized for global drylands. Adaptive capacity to these impacts will also require assessment at regional and local level.

Potential impacts in Africa

The potential impacts of climate change need to be approached at regional, national and local scales. In this discussion we cannot deal adequately with the diversity of global drylands, so Africa will be used as an example.

- ***Regional scale.*** Using some of the models employed by the IPCC in its Third Assessment (2001), a major study mapped climate change scenarios and poverty indicators over the entire continent computed in grid cells of 10 minutes of latitude and longitude, and based on predictions for the years 2020 and 2050.[90] Changes in the length of the growing period (LGP) were predicted using several models. The *most pessimistic* of them is reproduced in Figure 6, which is indicative only.[91] In the strongly seasonal climates of drylands, the number of days in which precipitation exceeds a critical minimum determines the growth and maturation of crops and forage. The LGP is thus defined agro-climatologically. According to this model, almost all of tropical Africa will experience shorter average growing periods by 2050, and in many areas by more than 20 percent. This defines a major adaptation challenge, and failure to adapt will have adverse impacts on livelihoods. More frequent crop or forage failures (also expected) must be compensated by changes in the crops or varieties grown, or in the movements of animals between grazing areas. Existing strategies – which are insufficiently understood by outsiders - will need to be supplemented by new knowledge.

It should be reaffirmed that the outcomes modelled in Figure 6 are only indicative.

- ***National scale.*** A study in Nigeria incorporated five major crops (maize, sorghum, millet, rice and cassava) and three projection periods (2010-39, 2040-69 and 2070-99).[92] Sorghum and millet are major food crops in the Nigerian drylands. Based on mean climatic conditions for 1961-90, a General Circulation Model predicted

[89] (*Ibid.*, p. 860 excluding North America, north Asia and Australasia)

[90] (Thornton *et al.*, 2006).

[91] The shortening growing period comes about as a result of the modelled increase in temperature not being offset by increased rainfall.

[92] (Adejuwon, 2006)

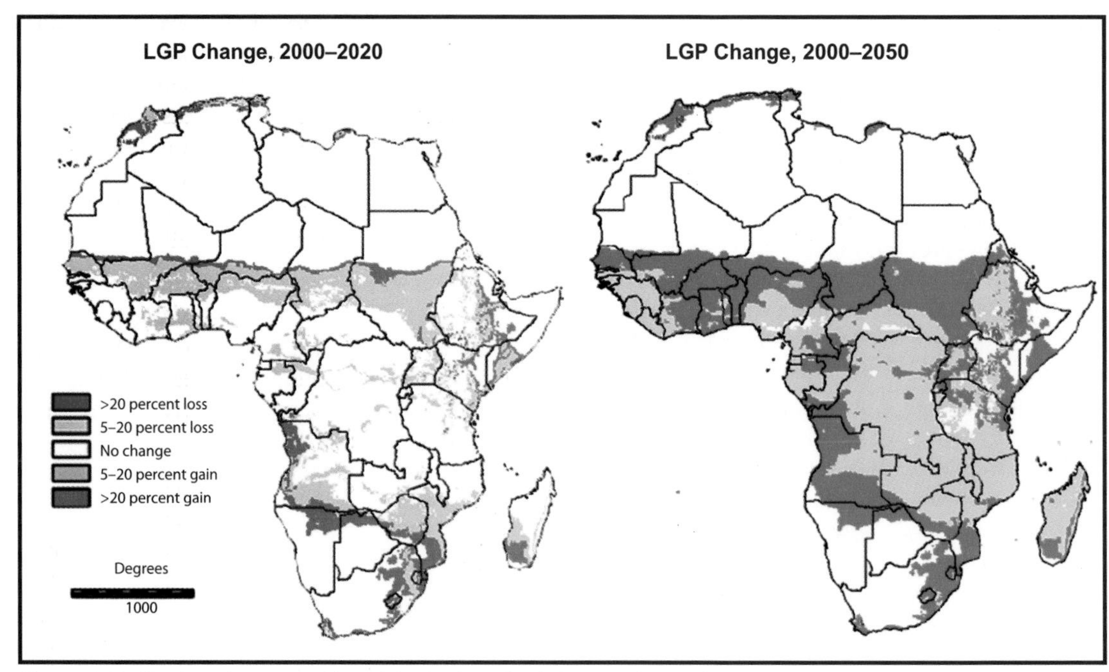

Figure 6. Indicative changes predicted in the length of the growing period in Africa.
Source: Thorton *et al.*, 2006.

potentially enhanced crop yields during the first half of the century and a decrease during the second half. Enhanced yields were explained by projected increases in rainfall, solar radiation, atmospheric humidity and CO_2 concentrations. Lower yields were explained in terms of continued global warming, resulting in maximum and minimum temperatures approaching the limits of tolerance for the modelled crops (except at higher altitudes). Moisture-based limiting factors would be replaced by temperature-based ones.

- ***Local scale.*** This is the scale at which adaptation must take place on a year-to-year and seasonal basis. Studies in the West African Sahel have recorded a range of indigenous knowledge about seasonal weather, including ways of forecasting the onset and ending of a coming rainy season.[93] Such knowledge can potentially guide adaptive actions on the part of farmers (choice of crops or varieties, timing of planting, use of fertilizers, strategies in the event of crop failure), or of herders (choice of grazing areas and herd movements). Such folk knowledge can be usefully supplemented by scientifically derived forecasts or probabilities. In semi-arid East Africa, farmers affirmed that they would significantly alter their strategies if given reliable seasonal forecasts.[94]

The adaptive challenge

The relations between crop growth and rainfall are not the result of simple cause-and-effect but rather reflect 'complex phases of growth and development responding to a climate that is multivariate, dynamic and heterogeneous'.[95] The same could be said of rangeland plants. In the same way, climate change and livelihoods will not be linked together in a simple cause-and-effect global relationship, as so often represented in popular media, but in interactive ways through mediating factors (such as access to land, water and grazing, income inequality, or gender). These factors have major importance in configuring the 'platform' on which adaptation is constructed. Too often it will be poor people whose adaptive capacities are the most constrained. Here are examples of complex

[93] (Ingram *et al.*, 2002)

[94] (Cooper *et al.*, 2008)

[95] (Adèjuwon, 2005)

challenges that call for adaptive response, whether at global, national or local levels:[96]

- Rapid demographic changes in response to climate impacts make resource management more problematic. For example, populations (whether human or animal) that migrate - as an adaptation strategy - can be a source of additional pressure on ecosystem services, when livestock temporarily concentrate at water points. Flexible administrative responses, stakeholder negotiations[97], and local natural resource agreements are needed.
- Ecosystems may suffer damage from habitat change, invasive species of plants or animals, biodiversity loss, or pollution when under pressure from land use change.[98] Adoption of ecosystem management approaches supported by empowering communities, sharing knowledge and stakeholder partnerships on a very large scale is necessary in drylands.
- Land scarcity and the diminution of landholdings, driven by a lack of alternatives to agriculture, with the effects of reduced rainfall or increasing temperatures on crop yields, may reduce farm incomes.[99] New economic opportunities must be sought, as well as technologies to enhance productivity.
- Disputed or insecure property rights, a lack of investment and subsequent soil erosion and degradation, may provoke out-migration with negative consequences for the human and ecological systems.[100] Securing of land rights is a recognized policy objective in some countries, though rights to grazing are not considered such a high priority.
- Global markets are difficult to enter, highly regulated and subject to quality control and source-tracing requirements that do not favour poor dryland producers.[101] Concern over carbon emissions and food miles increases downward pressure on food imports, affecting agriculturally dependent economies. International negotiations must play an increasing role in so-called 'free' markets.
- The HIV/AIDS pandemic reduces household labour, erodes assets, disrupts knowledge transmission and agricultural services.[102] The spread of climate-sensitive diseases alters with precipitation and temperature changes, leading to new disease burdens.[103] Public health systems must therefore be reinforced for rapid adaptation to new challenges.
- Threats of panzootics (e.g. avian influenza) attacking livelihoods and constraining trade may be compounded by an increased frequency of extreme weather events.[104] Readiness for pest outbreaks as well as for protecting human health needs to be increased.
- State fragility and armed conflict is current in some drylands.[105] Fragile states are ill-equipped to deal with climate change effects, so they may fuel such conflicts. Problems of resource availability and equity of access may be accentuated.[106] Much higher priority must be given to peace initiatives.

Conclusion

The adaptive challenge delineated above presents only one side of the picture. The other side consists in identifying present and potential capacities to adapt. Dryland peoples already manage climatic variability and uncertainty originating from non-climatic sources. A policy goal to strengthen such capacities converges on development goals. Thus, the greater the well-being (incomes, health, education, governance and life opportunities) enjoyed in a community, the greater its adaptive capacity. Sustainable management of environmental assets, including the maintenance of functioning ecosystem services as a buffer against climate change, is itself a critical form of adaptation. In the following chapters we attempt to delineate a new landscape of 'adaptive development'.

[96] This discussion draws on Morton, 2007; Anderson *et al.*, 2009.

[97] (Reardon and Vosti, 1992)

[98] (Easterling *et al.*, 2007; Safriel *et al.*, 2005).

[99] (Sadik, 1991; MEA, 2005)

[100] (Eriksen *et al.*, 2005; Lal, 2000; Vosti and Reardon, 1997)

[101] (Reardon *et al.*, 2003)

[102] (Barnett and Whiteside, 2002)

[103] (IRI, 2005)

[104] (ILRI, 2005)

[105] (FAO, 2005)

[106] (Eriksen *et al.*, 2008)

Children collecting firewood and seeds from *Acacia nilotica* in a forest close to a seasonal river in semi-arid area in Gedaref State, Sudan. © *Christopher Taylor*

CHAPTER 4
Realising the true value of ecosystem services

'A critical requirement of a one-planet economy is that economic calculations of all kinds take proper account of biodiversity and ecosystem services'.[107]

Dryland ecosystems have two characteristics that assist human communities not only to survive but to learn from nature. Ecological *adaptations* allow dryland plants and animals to reproduce, grow and survive in variable conditions. These include many species used by local people as part of their livelihoods. Indigenous trees of southern Africa, for example, have dozens of uses (food, beverage, medicinal, utilitarian, spiritual and cultural).[108] Domestic animals have been selectively bred for a very long time to adapt to local conditions, for example *Nguni* cattle (Box 8). Dryland ecosystems and species also have a dynamic *ability* to respond to low and variable rainfall and recurring drought in uniquely productive ways. Dryland ecosystems may be said to be *resilient* (Chapter 2).

It is important that sustainable use strategies are informed by an understanding of these adaptations and dynamics. This type of *knowledge* results from regular interaction between people and their environment. Research has shown that success can be attributed to social mechanisms embedded within communities for the transfer of knowledge and responses to environmental cues.[109] This knowledge also has a value, measurable not in monetary (market) terms but in the success or failure of household livelihood strategies over time.

Recognition of the true value of ecosystem services, and of the opportunities they offer, will enable better planning and realization of the full economic potential of dryland ecosystems, rebutting the common perception that drylands are 'economic wastelands'.[110]

According to the *Millennium Ecosystem Assessment*[111], ecosystems provide: (1) 'provisioning' services such as food, fibre, fuel, and water; (2) 'supporting' services such as biodiversity, soil formation, photosynthesis, primary production and nutrient, carbon and water cycling; (3) 'regulating' services such as air and water quality regulation and climate regulation and (4) 'cultural' services such as spiritual well-being through non-consumptive uses of the environment, and eco-tourism. Although many services do not enter markets, methods are available for estimating their value.[112] Moreover, the values of many lesser known commodities may be hidden in national economic planning, even though they feature in local and informal markets and contribute to the livelihoods of rural people.

The greater part of this chapter is concerned with the provisioning services of dryland ecosystems, and the need for them to be correctly valued in national accounting and policy making.

Box 8: *Nguni* cattle in South Africa

After being almost eliminated, the *Nguni* cattle breed is being revitalized for use by communal farmers in Eastern Cape Province. This hardy breed is known for an ability to withstand the environmental limitations, pests and cultural practices in this arid region. Unlike exotic breeds introduced during colonial times, *Nguni* are disease-resistant and productive in low-maintenance and low-input systems, such as those typical of poor communal farmers. They are highly prized for their beef and milk, skins and hides, draught power and manure which contribute to an integrated food security and livelihood strategy at a household level.

Sources: Musemwa *et al.*, 2008; Bester *et al.*, 2001; Jouet *et al.*, 1996; Mahamane 2001.

---

[107] (Adams and Jeanrenaud, 2008)

[108] (Sullivan and O'Regan, 2003)

[109] (Berkes *et al.*, 2000)

[110] (Dobie, 2001; Musemwa *et al.*, 2008)

[111] (MEA, 2005)

[112] (Barbier *et al.*, 2009)

Provisioning services: crop production

Although national statistics do not report separately on dryland regions, but merge data with that from more humid areas,[113] the contribution of drylands to national economies can be inferred from measures such as sector contributions to gross domestic product (GDP), per capita income, employment, public revenues, and export earnings. For example, agriculture contributed more than 30 percent of GDP in dryland countries such as Afghanistan, Burkina Faso, Kenya and Sudan in 2005, and over 20 percent in Chad and Pakistan.[114] In India, the arid and semi-arid tracts contribute over 45 percent of agricultural production, 53 percent of the total cropped area, 48 percent of the area under food crops and 68 percent of that under non-food crops; drylands account for nearly 80 percent of output of coarse cereals, 50 percent of maize, 65 percent of chickpea and pigeon pea, 81 percent of groundnut, 88 percent of soya beans and 50 percent of cotton. Moreover, because of the large extent of the drylands, a small rise in agricultural productivity has a large impact on the country as a whole.

These figures are direct values in terms of market prices. However, valuations may be made using other methods, which may take better account of the values of subsistence production to farming households, as well as indirect values such as those associated with family farming and aesthetic considerations ('cultural' services). These add to the total economic value of crop production by dryland smallholders.

Dryland ecosystems have low and variable rainfall and low biological productivity, and the achievement of food security is linked with crop productive capacity, food imports, market systems and growing populations. Despite these challenges, which require innovative and ingenious solutions to food insecurity, many dryland countries succeeded in maintaining food production per capita at constant or improving levels during the period 2000-2005 (Table 2).

In six West African countries having significantly large dryland regions,[115] food production per capita showed positive trends from 1977 to 1999, though with much inter-annual variability.[116] The cereal crops maize, millet, and sorghum dominate food production in these drylands, with rice in irrigated areas. Some of this additional output was achieved through extending the cultivated area, but it is significant that maize and millet yields per hectare remained stable (though low by world standards) or slowly improved. In Burkina Faso, yields of all four crops more than doubled over the period 1960-1999.[117] Rainfall was the primary determinant of yields from year to year. However,

Table 2. Food production in selected countries (per capita index, percent of 1999-2001 average).

Country	2000	2001	2002	2003	2004	2005
Bolivia	104.2	99.4	104.7	109.3	107.5	105.8
Botswana	98.8	106	105.4	98.4	99.6	99.3
China	100.2	102.7	107.4	110.1	114.8	117.8
Egypt	102.7	97.4	100.4	104.8	106.2	106
Ethiopia	98.4	105.6	106	100.5	101.6	100.1
India	99.1	100.8	94.9	100	99	97.8
Kenya	97.1	100.9	102	103.9	98.6	97.8
Namibia	95.8	103.7	109.1	123	122.2	121
Peru	101.3	101.5	106.1	107	104.3	106.2
Senegal	101.3	92.9	57.7	76.6	76.4	87.9
Tanzania	100.5	99.8	100.7	98.3	99.4	98.1

Source: FAO 2006 Statistical Yearbook.

[113] Exceptions are those countries that fall *entirely* within the drylands (e.g., Mauritania, Egypt, Yemen)

[114] World Resources Institute, http://earthtrends.wri.org

[115] Senegal, Mali, Ghana, Côte d'Ivoire, Niger, Nigeria (Mortimore. 2003; Toulmin and Guèye, 2003)

[116] FAOSTAT data, Food and Agriculture Organisation of the United Nations, 1960-2000.

[117] (Mazzucato and Niemeijer, 2000)

the long term trend was driven by growing demand from a doubling of the population between 1960 and 2000 and rapid urbanization. Structural adjustment policies introduced during the 1980s reversed an earlier declining trend. In eight countries, including six eastern African countries, food production increased throughout the period 1961-2002, albeit at a slow pace.[118]

This evidence demonstrates that the cultivable drylands play a critical role in ensuring national food sufficiency. The long-term trends are complex - demand and policy factors are important determinants, though hidden by annual variability in the rainfall.

Cultivation was extended to new land during this period. But output per hectare also showed rising trends in several of the West African countries, though fluctuating widely. This seems at odds with perceptions of declining soil fertility (Chapter 2), unless it can be accounted for by inorganic fertilization. But this is not affordable for many farmers. Uncertainty on this issue reflects the fact that the contribution of nutrient cycling and soil moisture – supporting services - to economic output is only rarely taken into account.[119] Yet sustainable management in future depends on such a valuation. We may estimate the market cost of the chemical soil amendments needed to 'restore' it to an assumed prior state (before 'mining' of nutrients began), but such an exercise has methodological limitations.

In drylands where agriculture is impossible (apart from under irrigation), food deficits must be met by importation. Such is the situation in several Middle Eastern countries. Recent initiatives to secure access to African drylands for large-scale Saudi Arabian farming operations are pointers to a future in which drylands will be seen as global, not merely local, assets.

Food emergencies have often dominated donor interest in drylands. However, policy should move on from meeting emergencies to supporting sustainable production, and linking producers and consumers through efficient and equitable markets.

118 (Holmén, 2005)

119 (ILRI, 2007)

120 (Hatfield and Davies, 2006)

121 Republic of Kenya, Statistical Yearbook, 2000, 2001, 2002.

122 (Hatfield and Davies, 2006)

Provisioning services: livestock

Fodder and water are the main (though not the only) provisioning services needed for keeping livestock in drylands. Fodder is obtained from natural pastures and from crop residues (in cultivated areas). Dry forests also provide browse for livestock (for example, an estimated 33 percent of feedstock requirements in the Sudan).

Direct values for livestock products are good indicators of value. The Chinese drylands (including Tibet) are home to 78 million cashmere goats which supply 65-75 percent of the world's cashmere fibre; and in Mongolia, pastoralism may provide 30 percent of GDP.[120] In Kenya, 50 percent of the national territory is too dry for farming, but is suitable for livestock. Over 60 percent of the national livestock herd is found there, providing 67 percent of the red meat consumed, 10 percent of GDP and 50 percent of agricultural GDP. The livestock sub-sector employs about 50 percent of the agricultural labour force.[121] Livestock provide 20-25 percent of agricultural GDP in Africa, and 25-30 percent in Asia.[122] In five West African countries, notwithstanding a doubling of the human population, FAO statistics show that the numbers of livestock units per capita remained constant or increased between 1961 and 2001.[123] In countries that depend on livestock for a large proportion of national income, such as Niger, the value of supporting rangeland ecosystems can easily be inferred. In the Sahel Region of Niger, on the border of the Sahara, livestock production contributes 46 percent of local household income.[124] In South America, Brazil and Argentina are among the top three world exporters of beef; 70 percent of Argentina's cattle and 13 percent of Brazil's are reared in dryland regions.[125] The global market for camel milk is estimated at USD 10 billion;[126] and in Ethiopia, leather exports provide 12 percent of national export earnings. Because dairy products are perishable, milk markets are localised unless cooled transportation facilities are available – which is rare. In tropical drylands, soured milk, butter and cheese markets respond to this constraint as well as providing valued inputs to nutrition. It would be perfectly possible to estimate total values for dairy production in pastoral countries.

123 (Mortimore, 2003)

124 (Zonon *et al.*, 2007)

125 (Linares-Palomino, 2009)

126 FAO Statistical Yearbook, 2006.

Box 9: Valorizing pastoral resources in Tatki, Senegal

In Senegal, since 2004 pastoralism has been recognised in policy as a mode of valorizing natural resources in arid rural areas. Tatki is in the northern Ferlo, a region with 200-250 mm average annual rainfall. Annual sales of livestock products were found to be worth USD 227 per capita, or USD 6,812 per (large) encampment, mainly consisting of sheep (60 percent by value, especially rams for festivals) and cattle (34 percent by value, especially draft animals). Of the labour used, 35 percent is hired, which is necessary because only 36 percent of the internal labour force were working with livestock at the time of the survey, the remainder either away seeking alternative incomes, or idle. The value of sales shows the system to be economically viable. Income diversification, however, is still regarded as desirable on account of the risky environment.

Source: Wane *et al.*, 2008.

Official statistics are based on market values – mainly the sale of animals – thus do not fully reflect the true value of pastoralism, which is usually the most profitable use of marginal lands, though policy is moving in this direction in some countries (Box 9). Nor do they necessarily recognise that the productivity of pastoral systems is often higher than that of alternatives. For example, in Africa it is 2 - 10 times higher per hectare than in ranching systems.[127] Neither are the values of cultural services, such as social coherence associated with keeping animals, factored into such valuations, particularly for mobile pastoralists.

This argument can be taken a step further: scattered and impermanent pasture resources are not seen merely as a constraint, to be managed through risk-averting herd mobility, but as an asset with its own particular configuration, offering not only risk but opportunity.[128]

Provisioning services: trees, energy and NTFPs

Where semi-arid or sub-humid dryland ecosystems support woody vegetation (especially in Africa and parts of South America), the value of trees and tree products is always critical to rural livelihoods. Trees contribute to national economies by providing fuelwood and charcoal for energy (for example, 80 percent of rural energy used in Mexico; 70 percent in Peru and north-east Brazil; 70 percent of national energy in the Sudan; and 74 percent of total energy consumption in Kenya, where charcoal is equal in value to horticultural products and only second to tea among marketed agricultural products).[129] A great part of this renewable energy is consumed by rural populations themselves, but urbanization and the growth of 'million cities' (Dakar, Khartoum, Kano) have generated large and growing markets, with geographically extended supply chains from remote sources.[130] Poor matching of demand and supply – especially in arid drylands where few trees are found – is responsible for large scale marketing and transportation. In most African drylands, wood fuel or charcoal continue to be cheaper and more reliable than higher technology alternatives, such as kerosene, gas or electricity. Wood requires a minimum of fixed capital (a 'three stone' stove), and may be bought a little at a time, and is thus accessible to the poorest people.

Following the oil price rise in the early 1970s, together with the Sahel Drought, a crisis narrative of desertification based on 'indiscriminate deforestation' became popular among environmental authors, development agencies and NGOs.[131] Much effort was spent on estimating 'wood fuel gaps' between demand and supply, from which predictions could be made of when woodland resources would be exhausted. These estimates appeared to justify large-scale surveys of supply and demand as a basis for more sustainable, state-directed forest management, large investments in plantations, and policies to promote alternative energy sources.[132]

[127] (Scoones, 1994)

[128] (Krätli, 2008)

[129] (ILRI, 2007; Linares-Palomino 2009)

[130] (Cline-Cole *et al.*, 1990)

[131] (Eckholm *et al.*, 1984)

[132] For example, in Nigeria: (Cline-Cole, 1998; Hyman, 1993)

However with democratization, forest policies have tended to move instead in the opposite direction, especially as decentralization and participation have been shown to work better in rural areas than the embattled policing of diminishing forest resources by the state.[133] Also some estimates of 'wood fuel gaps' around urban areas have been shown to be exaggerated,[134] as the treeless deserts predicted in the hinterlands of major cities have failed to emerge, at least in West Africa. In-depth research has shown that rates of household consumption are likely to decline with increasing price, especially in urban areas.[135]

Ownership is critical. After several decades, Tanzanian villagers were given the right to choose the best project approach to sustainable forest management, and revived their *ngitili* institution for managing private or communal woodlands. Implementation of this approach added value to farming and provided significant alternative incomes, exceeding expectations in uptake (Box 10).

Box 10: *Ngitili* forest and grazing reserves in Sukumaland, Tanzania

A culturally established practice among the Sukuma, the *ngitili* is either a private or a communal grazing and fodder reserve, supported by a revised forest policy which places a strong emphasis on participatory management and decentralisation. The *ngitili* provides dry season forage, fuel and poles, medicinal plants, wild fruits and other foods (especially during food shortages), shade and quiet. As a result of a forest conservation project in Shinyanga region, which supported boundary mapping and title deeds, the institution was revived after years of neglect, and the total area increased from 78,000 to >300,000 hectares, in 833 villages. *Ngitili* provided an average of about USD 1,000/family/year. In addition, soil and biodiversity conservation values were identified by the people. However, better off people earned four times as much per capita as the poorest. *Ngitili* experience shows the value of a customary resource management system in mobilising local knowledge against degradation and of social institutions to implement improvements and changes, as well as for regulating access to grazing.

Sources: Barrow, 1996; Ghazi *et al.*, 2005.

Based on work in West Africa, a transition model offers a basis for forest policy that recognises indigenous knowledge and practice as well as the economics of supply and demand. On the settlement of new lands, the clearance of dry forest dominates land use decisions made in a context of scarce labour and abundant land. In Maradi Region, Niger, for example, cultivated land increased from 59 percent of the total area in 1975 to 73 percent in 1996.[136] This was the culmination of several decades of northward migration by Hausa farmers. On privately owned land, trees that regenerate naturally - and sometimes are planted - are valued for multiple purposes, including comestible fruit, fibre, fodder, medicines and timber. As the population increases, markets for these products grow. Timber is normally harvested only from dead or branch wood. Inventories of parkland species show a northward trend in introductions from Sudanian to drier Sahelian zones and from Guinean to Sudanian, taking advantage of wetter decades before the 1960s.[137] The emergence of 'farmed parkland' has proved durable even during droughts when food shortages have driven people to search for alternative incomes, including selling firewood.[138] Such transitions to sustainable tree management on private farms are driven by increasing land and product values. In Maradi, the protection of naturally regenerating, indigenous trees (*défrichement amelioré*) was promoted by an NGO and by a donor-funded development project, with visible impact on the landscape (Box 11).[139] Recent satellite imagery shows accelerated uptake of this land use strategy probably reflecting a mix of improved rainfall, changing economic incentives, more secure right-holding, and successful extension.[140]

In Mexico, north-east Brazil and Peru, fuelwood provides more than 70 percent of rural residential energy; and in Peru, for example, the carob tree (*Prosopis juliflora* and *P. pallida*) is a very important source of domestic fuel, due to its excellent calorific properties: it burns evenly and hot, the wood does nor spit, spark or smoke excessively, and the smoke is never unpleasant. It must be noted, however, that its popularity is also linked

[133] (Ribot, 1995)

[134] (Foley, 2001)

[135] (Cline-Cole *et al.*, 1990)

[136] (Mortimore *et al.*, 2005)

[137] (Maranz, 2009)

[138] For example, in the Kano region, Nigeria (Cline-Cole *et al.*, 1990)

[139] (Jouet *et al.*, 1996)

[140] (WRI *et al.*, 2005)

Irrigated agroforestry site in Kassala State in semi-arid to arid North-Eastern Sudan. © *Caterina Wolfangel*

Box 11: Tree regeneration on farms in Niger

When farmers migrated northwards during the early and mid- twentieth century, in response to growing population and new markets, they cleared the natural woodlands to make way for crops of millet, groundnuts and cowpeas. Assisted by frequent wet years, the cultivated area increased from <35 percent in the 1950s to 59 percent in 1975 and 73 percent in 1996. By the 1980s, trees were sparsely distributed, and soil fertility loss, wind erosion and increased risk from droughts were reported. A scarcity of fuelwood, timber and valuable NTFPs had developed as forests became degraded. Food security was low and food aid was required often. Development interventions based on tree-planting (often exotic species) had disappointing impact. But beginning in the 1980s, an NGO-led programme promoted on-farm protection of naturally regenerating indigenous trees – an activity already well established in long-settled areas of the Sahel (such as Kano, Nigeria and the *basin arachide* of Senegal). This was taken up by a state- and donor-sponsored development programme and soon became accepted as good practice in the Maradi and Zinder Regions. A recent survey claims that 5 million ha of land and 4.5 million people are now enjoying the new trees and their beneficial effects on soil fertility, erosion control, and risk reduction.

Sources: Raynaut, 1980; Jouet *et al.*, 1996; Mahamane, 2001; WRI, 2008.

to its ubiquity. Where trees are present they often grow in large numbers and on common land, thus freely available to all sections of society.[141]

Non-timber forest products (NTFPs) can be valued along with other ecosystem services integrated into national accounts, policy frameworks and local decision making. They can also point to policy choices available, and provide valuation guidance for assessing corporate performance and ecological footprints.[142] Table 3 summarises the findings of a study in the Kgalagadi South District of Botswana, using valuation methods to quantify the benefits of selected products at the household and community levels. Here, in a district where many live on less than a dollar a day, plant- and livestock-based activities are valued at USD 1,394 per year at the household level, and community enterprises achieved on average USD 3,590 from hunting and USD 8,735 from tourism per year. Total estimated value to the District is USD 191,256, and estimated asset values are USD 985,800.

Table 3. Values of ecosystem services in Kgalagadi South District, Botswana (in USD)

	Direct Use: Annual profits of enterprise[1]	Direct Use: District total [2]	Asset Value[3]	Indirect Use
Plant Use	270[4]	91, 874	599,430	-
Livestock Use	1, 124[4]	68, 216	Nil	-
Trophy Hunting	3,590[5]	7,739	27, 030	-
Tourism	8,735[5]	23,427	369,340	-
Total	-	**191, 256**	**985,800**	-
Carbon Sequestration	-	-	-	111, 300
Erosion Protection	-	-	-	68, 400

Adapted from Madzwamuse *et al.*, 2007.

1 Private values (net annual private profits to investment realised by households or community enterprises), as expressed in monetary or in-kind transactions.
2 Economic values or estimated contribution to national income (outputs less the costs of production).
3 Present value of expected future contribution of dryland ecosystems in terms of economic rent.
4 Per household
5 Per community enterprise

141 (Linares-Palomino, 2009)

142 (Adams and Jeanrenaud, 2008, p. 68)

Aloe vera plant (*Aloe vera, Aloe barbadensis*) and Aloe vera cream © *P. Royer / Still Pictures*

Box 12 illustrates both the opportunities and the barriers in NTFP exploitation in Latin America.

In Senegal, the sales of NTFPs (harvested fruit, leaves, seeds, gum, roots, bark, honey) in Kolda and Tambacounda Regions of Senegal in the year 2000 were worth USD 2 million, and the value added along the supply chain averaged 48 percent; the value added to game by-products reached 63 percent.[143] Extrapolated to national level, including value added to urban markets, a median estimate of the annual economic contribution of NTFPs was USD 6.3 million. This is equivalent to an addition of 14 percent to conventional estimates of value added in the forest sector (timber, fuelwood and charcoal). Fresh water fisheries, based on studies in two of the three major river basins, were estimated to be worth USD 14.5-19.6 million in value added in the country as a whole. These values were 19-26 percent of the value of marine fisheries, the primary sector by value in the Senegalese economy. If recent movements in the value of the USD are taken into account, the national estimates increase to USD 8.4 million for NTFPs, and USD 19-26 million for freshwater fisheries.[144] In sum, between 19 and 35 million USD of value added from

Box 12: NTFPs in Mexico, Ecuador and Bolivia

Non-timber forest products play an important role in the economy of rural societies in the Latin-American drylands. Ethnobotanic studies have shown the importance and diversity of useful wild plants in the region. However, few of these uses end up in economic networks and there is little information about economic valuation studies. In Chamela, as many as 162 plants species are, or have been, used for medicine, timber, wood fuels, materials, food, beverages, and spices. They are commercialized in local, regional, national, and international markets. For instance, precious timber species such as *Cordia* spp., *Enterolobium cyclocarpum*, *Tabebuia* spp., and *Pirhanea mexicana* are commercially extracted. A study of wild plants from southern Ecuador, a region with several subtypes of drylands, reported 354 edible plant species used by the local people, but only 21 of those were actually sold at local or regional markets. NTFPs benefit the poor, the less poor and women; they are covered by little policy or legislation; market information is inadequate; but where successful there is a risk of unsustainable use.

Sources: Maass, 2005; Marshall *et al.*, 2006; Van den Eynden *et al.*, 2003; Linares-Palomino, 2009.

[143] Ba *et al.*, 2006)

[144] Based on the exchange rate of USD 1.00=F CFA 463.446 (10/09/08).

wild products are currently excluded from national accounts. At a minimum, this would represent 10 percent of the annual GDP recorded for Senegal (approximately 20.6 billion USD) in 2007.[145] Similarly, the Sudanese forests are famous for Gum Arabic, which earns 14 percent of annual export income.[146]

African studies point to an under-valuation of ecosystem goods and services in national accounts which impedes planning on the basis of the true potentials of drylands. Given prevailing attitudes of governments and donors, the problem may likely extend beyond Africa.

Supporting services

The value of supporting services is subsumed by, and implicit in the provisioning services that they enable. For example, cycling of soil nutrients, moisture and biological agents is critical to the productivity of the soils under cultivation, pasture, or natural vegetation. Biodiversity may be improved or reduced by grazing management. Although formerly blamed for desertification, grazing and animal impact can stimulate pasture growth, reduce invasive weeds and may improve mulching, and mineral and water cycling.[147] Rangeland health and integrity are better where mobile pastoralism is practised.[148] This allows recovery after grazing cycles and seed propagation.

Agro-diversity supports adaptive and flexible agriculture in risk-prone environments. Diversity among wild and domesticated animals furthers efficient use of food resources. Extensive farming is usually assumed to reduce biodiversity. However, as farming intensifies, greater priority may be given to preserving it as trees, shrubs and herbs have food, fodder, medicinal and other values. In one village in the western Sahel, 135 useful species were recorded, and attitudes among the population were found to be strongly conservationist.[149] But estimating values of these ecosystem services separately from those of the provisioning services they support is beyond the scope of this study.

Regulating services

The regulating services provided by dryland ecosystems are critical for their management by dryland communities. But they are the most difficult services to value, and are often poorly understood.[150]

Water holding. There are claims that effective pasture management can improve infiltration of water, reduce run off, and thereby raise water tables. If each millimetre of additional rainfall captured represents 1 litre more usable water per m^2, or 1,000,000 litres more water per km^2, it is worth investigating the positive role of pastoralism.[151] Dry farming and irrigation both have an impact on sub-surface water, the first through infiltration effects and the second through withdrawal. The values of water services are immediately apparent when mismanagement leads to soil desiccation or salinization.

Soil fertility. Nutrients, biological organisms and physical properties of soils are critical to supporting ecosystem services for farming and grazing. Pastoralism does not necessarily generate overgrazing and land degradation, because collective action or institutions regulating access can ensure sustainable use. Attempts to manage the ecosystem through 'controlled grazing' may

Box 13: Controlled grazing schemes in Senegal

In the early 80's, the German GTZ collaborated with the Senegalese Forest and Water Service to create a model to test a new way of managing rangeland and herds around the borehole of Widou Thiengoli in northern Senegal. The model was based on trying to find the right balance between the number of cattle and amount of fodder available. In order to do this, the project provided special benefits to those few families who were allowed to use the pasture and water enclosed and protected by barbed wire fencing. But selling off animals soon after weaning (as planned) was not profitable enough. Animals which had gained advantage in good years within the enclosure were at a disadvantage in years of poor rainfall. In wet years, insufficient trampling of forage and soils led to the disappearance of preferred grasses. Fencing some families in, and others out, of what had once been a common resource enabled the elite to capture the benefits. Families within who benefited during good years found themselves rejected by others in the bad years when they had no choice but to cut the wire and let their animals venture out onto the common range.

Source: Thébaud *et al.,* 1995.

[145] (CIA, 2008)

[146] (Bennett, 2006)

[147] (Hatfield and Davies, 2006, p. 22)

[148] (Niamir-Fuller, 1999)

[149] Mohammed, S. In: (Mortimore *et al.*, 2008)

[150] (Barbier *et al.*, 2009)

[151] (Hatfield and Davies, 2006, p. 22)

not succeed (Box 13). Pastoralism can play an important role in maintaining ecosystem health and resilience, promoting water and mineral cycling, and protecting biodiversity. Paradoxically, under-grazing can lead to encroachment by unwanted trees and shrubs - a major issue in ecosystem management in southern and eastern Africa.

Farmers in the Kano region of Nigeria have stabilised soil fertility, though at rather low levels, even under annual cropping regimes. Fertility is now a function of management. Organic matter recycles chemical nutrients and biological organisms and the fertility varies sharply between and within plots, depending on a farmers' access to animals, compost, weeds and chemical fertilizers if affordable.[152] In such a densely-populated farming system, there is a close tie between poverty reduction and the capacity to manage ecosystems sustainably.

Carbon. Estimations of carbon stocks, sequestration and values vary from one ecosystem to another, and among sources, as would be expected. Grasslands (which include some but not all drylands) store approximately 34 percent of the global stock of CO_2. If the sink values of major biomes are estimated separately, the tropical savanna and grasslands average 0.14 tC/ha/yr, compared with only 0.01 tC/ha/yr for cropland (which is assumed to include both tropical and temperate). The corresponding values for Net Primary Productivity are 7.2 and 3.1 tC/ha/yr.[153] Such figures should be treated as illustrative.[154]

Sound management can promote carbon sequestration, erosion limitation, water storage (and purification), and nutrient (including carbon) cycling. Radiation absorption in grasslands can mitigate the effects of drought or erratic weather patterns. In Table 3 (above), for example, 'indirect' economic values are assigned to carbon sequestration and erosion protection services in the Kalahari.

Regulating services enhance livelihood opportunities and reduce vulnerability to the impact of climate change, notwithstanding the extreme and unpredictable elements of these landscapes and their limited access to water.[155] The Chamela ecosystem in Mexico provides a good example (Box 15).

Management makes a difference to what the regulating services can deliver, not only to dryland people, but to other ecosystems. For example, The costs of carbon sequestration vary according to management practice, and a case-specific approach is necessary (Box 14). Carbon capture may help to stabilise the global climate system, and at the continental scale, the export of atmospheric particulates (Saharan dust) across the Atlantic Ocean to the tropical forests of South America and the Caribbean contributes soil phosphorus on a scale that is significant in the long term.

Box 14: Impact of management on Carbon sinks and stocks in Latin America

Land degradation is a major cause of poverty in many parts of Latin America. Results of a five year on-farm research project in Colombia and Costa Rica show that compared to degraded pastures, cultivation of perennial grasses and other good management practices can significantly increase soil C stocks within short periods of time. Average annual sequestration rates across all sites and practices were >4 tC/ha/year. Inclusion of dispersed trees in pastures further increases total C stocks without significantly affecting livestock productivity.

Source: t'Mannetjie *et al.*, 2008.

Cultural services: tourism

Tourists are attracted to drylands by wildlife, scenic beauty, and cultural artefacts including ways of life of pastoralists in particular. Some indication of the importance of tourism to some drylands is given in Table 4, which reports the numbers of tourists and tourist revenues for selected African countries.

Tourists' interest in wildlife has led to parks, reservations and conservation projects having conflicts of interest with rapidly expanding agricultural or ranching interests. Although tourism brings substantial revenues to the national economies of many dryland countries (Kenya, where tourism accounts for 13 percent of GDP, is, however, exceptional[156]), the benefits do not necessarily flow to dryland farming or pastoral

[152] (Harris, 1998)

[153] (Grace *et al*,. 2006, cited in Tennigkeit and Wilkes, 2008)

[154] It is possible that the estimated areas of tropical savanna and grassland include some tropical cropland, thus invalidating the comparison.

[155] (Burton, 2001)

[156] (Hatfield and Davies, 2006: p. 21)

Box 15: Regulating services in the ecosystem of Chamela, western Mexico

Climate regulation. Dryland forests provide shade and moisture to farmers and their animals. At regional scale, changes in albedo as a result of large-scale forest transformation can significantly modify the relative importance of the sensible and latent heat fluxes, changing regional energy and water budgets. Dry forest landscapes in Mexico store carbon at about the same rate as evergreen forests, but emissions from the burning of biomass may be higher.

Soil fertility maintenance. The forest has evolved tight recycling mechanisms to avoid nutrient loss from the system, including a dense leaf litter layer, microbial immobilization of nutrients during the dry season, nutrient resorption prior to leaf abscission, forest resistance to fires, and high soil aggregate stability. When the forest is transformed, these fertility maintenance mechanisms are weakened.

Flood control. The region is exposed to highly erosive storms. But there is always a leaf litter on the forest floor that protects the soil, keeps high infiltration rates, reduces runoff and erosion, and floods. When the forest is transformed into agriculture and pasture fields, soil cover decreases and infiltration rates diminish, resulting in higher rates of erosion and sediment transport downstream.

Bio-regulation. The presence of native and introduced pollinators is needed by many of the crops that, in 2000, accounted for USD 12 million. Vertebrates, such as bats, are essential pollinators of wild and domesticated species of cactus and agave, as well as trees of the family Bombacaceae.

Sources: (Linares-Palomino, 2009, citing others).

households displaced or constrained by parks. To capture a larger fraction of the USD 6 million tourism industry, Tanzanian pastoralists are now starting locally owned facilities with donors' help.

Approaches based on the engagement and participation of local communities in co-managing protected areas (e.g., CAMPFIRE in southern Africa) have made slow - though significant - progress. The clash of interests is exemplified in Kenya, where between 1977-78 and 1994-96, wildlife decreased by 61 percent, only increasing in one of 24 districts, while livestock also decreased, but only by 30 percent. Cultivated land, on the other hand, increased from 1985 to 2003, with that planted with maize from 1.2 million ha to 1.6 million ha, and that with beans from 0.6 to 0.9 million ha.[157] The result is a mosaic of landscapes: those transformed by land use change, and those energetically protected for tourists.

Table 4. Numbers of tourists and value of tourism in African countries with drylands.

	international inbound tourists[1]			international tourists receipts[2]			
	Thousands			$ millions		percent of Exports[3]	
Country	2000	2006	percent change	2000	2006	2000	2006
Southern Africa							
Botswana	845	1675	98	234	539	7.7	10.2
Namibia	614	833	36	288	473	17.9	29.6
West Africa							
Senegal	369	769	108	166	334	11.5	13.2
Eastern Africa							
Kenya	943	1536	63	304	1182	11.3	19.8
Tanzania	459	622	36	739	950	57.7	29.6
Ethiopia	125	290	132	24	639	2.4	29.1

Source: World Bank. 2008, 2002 World Development Indicators.

1 The number of international inbound tourists is the number of visitors travelling to a given country for purposes other than business.
2 International tourists' receipts include prepayments for goods or services.
3 The share of receipts in exports is calculated as a ratio of goods and services to exports.

157 (ILRI, 2007: pp. 3, 74)

Tourist receipts are vulnerable to events that undermine perceptions of personal security. Before the recent onset of armed conflict in the Sahel Region of Niger, the Aïr and Ténéré nature and biosphere reserves were estimated to generate tourism revenues worth about USD 6 million.[158] Tourists tend to converge on established attractions and famous countries or regions, and so the benefits of tourism are unevenly distributed in space as well as fragile. This is not a developmental strategy that is replicable in all drylands.

Other cultural or spiritual services are not considered here, although they are real and significant to their beneficiaries.

Re-evaluating ecosystem management

In the drylands of poor countries the dominant systems of land use may be simplified as follows:

- a. Mobile herding of livestock ('nomadic pastoralism')
- b1. Extensive rain-fed farming with semi-sedentary livestock herding
- b2. Intensifying rain-fed farming with integrated livestock keeping
- b3. Small-scale irrigated farming in river valleys and local depressions

Herders and farmers have been accused of causing land degradation through over-grazing, over-cultivation, and deforestation. However, decades of unsuccessful attempts to transform them have forced a re-evaluation of these systems. Development agencies have scaled down their expectations and field studies have improved scientific understanding of their adaptive strengths as well as their vulnerabilities. For example, mobile pastoral systems are found to be compatible with biodiversity conservation and sustainable ecosystem services.

The mobile livestock herding systems (a, above) were once a target for cattle ranching or 'controlled grazing' schemes. These were tied to the idea of 'carrying capacity' – the largest number of weighted animal units supportable through low rainfall years – in a bounded area. However, it was shown conclusively that mobile herding is more productive than ranching because it permits better use to be made of feed resources that are highly variable in time and space.[159] African evidence indicates that such opportunistic grazing systems give better economic returns per ha than livestock reared under ranching conditions.[160] Controlled grazing cannot adjust adequately to this variability (Box 13). But the WoDaaBe (and others) can.

In the northern (drier) Sahel, although they are less abundant than those in the south, pastures have been shown to be richer in some elements.[161] This is well known to pastoral communities, who regularly move their animals north during the rainy season to fatten their livestock in preparation for the difficult dry season. Such transhumance is rewarded. In Niger, nomadic cattle are 20 percent more productive than sedentary cattle in terms of annual reproduction, levels of calf mortality, and annual milk production.[162]

Commercial crop farming has remained secondary to small family farms in most drylands. Even on land formerly reserved for cattle ranching, small-scale cultivation has recently been introduced (e.g., in Kenya). Such farming is risky and livelihoods need backing up with income diversification strategies. The primary charge levelled at small-scale farming in drylands is the destruction of soil fertility, either through exposure to erosion, or through nutrient and organic matter depletion from repeated cropping with inadequate replacement inputs. Fallow periods become shorter under conditions of increasing demand for land and divisible inheritance. Full vegetative recovery is frustrated, and the redistribution of nutrients from fallow to field via grazing livestock is reduced.

The wisdom of rotating and mixing crops (including nitrogen-fixing legumes), recycling crop residues and weeds as highly nutritious fodder, and maintaining livestock on farms as natural rangeland diminishes, is now recognised and the thrust of agricultural extension to small-scale farmers is towards supporting such intensifying systems rather than

[158] (Zonon *et al.*, 2007).

[159] (Behnke *et al.*, 1993; Sandford 1983)

[160] (Scoones, 1994; Western, 1982)

[161] (Breman and de Wit, 1983)

[162] (De Verdière, *pers. comm.*)

transforming them according to commercial models. Food security is given official recognition rather than being dismissed as mere 'subsistence' – an early stage in development. Nevertheless, fertility maintenance is a major issue despite the remarkable persistence of resource-poor farming systems, for example, maintaining 100 years or more of annual cropping in the Kano Close-Settled Zone of Nigeria.[163]

In reality, the systems 2a and 2b (above) merge imperceptibly as a growing scarcity of land forces an increase in labour and other inputs. Financial resources are increasingly critical as inorganic fertilizers come to be seen (and are promoted as) a solution.

Also in the Sahel, ambitious re-afforestation programmes were supported by governments and donors as a solution to 'indiscriminate deforestation' and the perceptions of degradation in farming areas during the drought cycles of the 1970s and 1980s. Based on exotic, fast-growing species, they were largely unsuccessful, and public investments in nurseries, water supply, distribution and planting were lost. Meanwhile, farmers in intensifying systems have long practised tree planting and the protection of natural regeneration on their farms.[164]

Some reasons why existing practice was systematically under-valued for so long in Africa include the following:

- Smallholders' motivation to maintain viable farms or herds for their heirs was under-estimated, in face of the myth that poor people always have short-term planning horizons.
- Local knowledge was not adequately appreciated by development agents, nor were local capacity to evaluate new (external) knowledge, exchanges from farmer to farmer, and 'spontaneous' experimentation.
- Livelihood goals – which include health, education, welfare, income diversification and migration – complicate the uptake of new practices in natural resources management.
- Development practice was based on an equilibrium model, whereas local practice sought to adapt to uncertainty (or dis-equilibrium).

Some of this diagnosis may apply also to drylands outside Africa. Now, however, the values of low external input production systems in drylands are potentially transformed by the crisis in global sustainability and climate change. Having for long been an 'investment desert'. the drylands should now profit from having among the lowest carbon footprints in the inhabited world. We can no longer afford to treat local knowledge and practice as 'conservative', 'backward', outside the market, and necessarily destructive. Not only can rural drylands boast low carbon emissions, but efforts are being made to recruit their help in mitigating global warming. Agricultural intensification (based on labour, skills, and organic cycling), tree protection and planting, and withdrawal of cultivation may not only reverse degradation but are now being promoted in schemes to sequester carbon. Financial compensation for these environmental services may support the local economy, but the social outcomes of such schemes are controversial, especially in Latin America. So are the risks posed by new commercial interests in biofuels or food production for export (see Chapter 7).

Conclusion

Evidence has been given to support a re-evaluation of ecosystem services and management in drylands. This will enable governments and donors to include dryland goods and services in national accounts. In conjunction, a re-evaluation of local management systems should involve a change in the attitudes reflected in public policy. These re-evaluations make a case for improving incentives for public investment in drylands. These issues are taken up in the next chapter.

[163] (Mortimore, 1993a; Mortimore, 1993b)

[164] (Cline-Cole *et al.*, 1990)

CHAPTER 5
Investing in drylands

Drylands in poor countries are 'investment deserts', except where valuable minerals have attracted inward (and short-term) investment. Because of their risky climates and low bio-productivity they need inward financial flows if they are to achieve their potentials. But dryland regions in countries such as Argentina, Australia, Israel, and the USA stand in sharp contrast, having benefited from higher capital investments. Their relatively advanced development provides the strongest evidence that drylands need not be poor. Investors in poor countries have, however, preferred the high potential regions.

Drylands in tropical Africa have tended to be graveyards for well-intentioned project investments. For example, in northern Nigeria a large-scale environmental afforestation scheme was embarked on by the Federal Government in areas considered to be most at risk from desertification in 1977-78. Seedlings of fast-growing exotic trees were multiplied in a network of nurseries (supported by boreholes, pumps and water tanks), free distribution in composted containers was undertaken, shelter belts set aside, staff hired, transport equipment and technical and extension services provided. After five years there was little left of this project. Seedling establishment in the shelter belts was very low, owing to late planting or droughts, browsing animals broke down the fences, free seedlings were not watered after planting, financial and staff resources were soon inadequate, and no permanent impact on the landscape was achieved.[165] It is tempting to see drylands in poor countries as incapable of yielding a good economic return. Such a characterisation defines the prevailing poverty of drylands just as moisture constraints define their bio-productivity. But this is not the whole picture.

Landscape investments

The behaviour of small-scale farmers in some African drylands challenges the stereotype of unacceptably low returns to investment. Recent studies show that their long-term investment strategies, unrecorded and so usually ignored in macro-economic planning, have gradually transformed some densely-populated farming landscapes. Often constrained by poverty, smallholders invest incrementally, and many of their investments are created by labour. Finance is sourced from off-farm incomes as well as agricultural profits. It is highly significant that many of their strategies are designed to conserve the productive capacity of their land, rather than 'mining' it as often alleged by outsiders. Among well-documented cases is Machakos District in Kenya (see Box 16).

Box 16: Smallholder investments in Machakos/ Makueni Districts, Kenya

In a study of landscape management, 1930-1990, the following investments were found to be made by virtually 100 percent of farmers in the districts:

- Clearance and enclosure of farm land
- Improved management of enclosed pastures
- Building of soil and water conservation structures
- Adoption of new technologies
- Integration of crop and livestock production
- Planting and protecting economic trees on farms
- Purchase of organic and inorganic fertilizers
- Purchase of improved seeds
- Erection of grain stores, poultry houses, and livestock bomas
- Acquisition and hire of farm transport vehicles
- Building, improving and extending farm houses.
- Purchase of animals, equipment, immunisation, salt cures

Source: Tiffen *et al.*, 1994.

[165] Source: unpublished material and field work

While some of these improvements were promoted through government- and donor-funded project interventions and credit schemes, it is notable that these were short-lived whereas the landscape transformation is long-term. Its sustained momentum is due to a positive social and economic evaluation of sustainable ecosystem management driving the development process. However, such incremental investment is an untidy process from an economist's perspective. Taken forward in years of prosperity, it may regress in years when food is scarce and resources must be diverted to consumption. At all times, land management investments compete with other livelihood priorities. Farming households try to access incomes outside agriculture to boost their resources. Nevertheless, productivity per hectare of farmland has been increased (or its decline averted), land values have risen, and markets for land, labour and skills have grown. New crops and new livestock activities (e.g., fattening) have developed.

In the central Plateau of Burkina Faso, small-scale investments in soil and water conservation accomplished a turn-around in agricultural productivity between the 1980s and 2000, with improved crop yields and other benefits, even reversing the trend of rural-urban migration.[166] In eastern Burkina, the agricultural landscape has been perceptibly changed through intensification of crop production by small farmers.[167] In the so-called 'Peanut basin' of Senegal, although crop production was damaged by a decline in the groundnut sector and low rainfall adversely affected grain yields, farmers invested in livestock in response to buoyant prices.[168] In the Kano Close-Settled Zone of northern Nigeria, a farming population living at more than 200/km^2 maintains one of the most intensively farmed landscapes in Africa, despite having an average annual rainfall of less than 700 mm.[169]

Landscape transformation is an indicator of agricultural intensification when it is based on labour, local knowledge, efficient nutrient cycling, and the use of organic inputs in combination with an affordable minimum of chemical fertilizers. Such landscapes are spreading rapidly outwards from their original nuclei (often in the vicinity of towns), driven by growing rural populations, new and growing urban markets, and increasing demand for, and values of, cultivable land and multi-purpose trees. In northern Nigeria and southern Niger, such market expansion has been found to have a beneficial impact on the ecosystems, pushing them towards more sustainable trajectories, in contrast to the wilderness of soil degradation predicted in some scenarios.[170] However, a recurring theme in analyses of intensifying systems is the diversity of livelihood circumstances and priorities. These, together with variable agro-ecological conditions, caution against generalisation and predetermined investment targets (which are unfortunately insisted on by many donor-funded projects).

Elsewhere in Africa and outside it, increasing pressure of demand on natural resources likewise pushes production systems towards maximising the efficiency of the scarcest factor. In the pastoral zone where rainfed farming is impossible, labour or skills may be the limiting factor. There may be less opportunity to invest in pastoral systems, where little effort has gone into researching new breeds or technologies compared with that spent on crop breeding and agronomy for farming systems. Where large-scale management is the norm, or where out-migration has significantly reduced the labour force, investments of financial capital must substitute for labour. Because capital is easily moved elsewhere, there is a greater risk of unsustainable practices damaging the ecosystems.

Outside Africa

At a general level, landscape transition is under way in most drylands, its pace depending on the level of investment, much of it at a micro-scale. Rather than the over-simplified narratives of degradation popular with many policy makers and media channels, a more productive approach is to begin with an understanding of dryland landscapes based on investment in a transition to sustainable management under changing conditions. Evidence that this is already happening in many systems should be taken seriously as it offers a platform on which policy and interventions may be built.

[166] (Reij and Thiombiano, 2003)

[167] (Mazzucato and Niemeijer, 2000)

[168] (Faye *et al.*, 2001)

[169] (Harris and Yusuf, 2001; Harris, 1998; Mortimore and Harris, 2005)

[170] (Ariyo *et al.*, 2001; Mustapha and Meagher, 2000)

Public investments

The 'investment desert' - as noticed above - cannot bear fruit without investment and much of this must come from external sources - government, donors, NGOs and the 'private sector'. This is because for decades, drylands have exported their human, social and financial capital to urban areas or regions offering a less erratic return. The public sector in particular must accept this in its rationale for investment in drylands.

Evidence from India and China indicates that economic rates of return to public investments may be higher in rainfed dryland regions than in irrigated and more humid regions. In India, rural districts were classified into predominantly irrigated or rainfed, and the rainfed areas were subdivided into agro-ecological zones, including semi-arid. Five categories of public investment were analysed: research on high-yielding crops, rural roads, canal irrigation, electricity provision, and education. There is considerable variability among the rainfed zones, but in roads, electricity and education, the semi-arid zones performed better on average than the irrigated areas, and the investments had a greater impact in reducing the numbers of poor people.[171] Comparable results were obtained in China.[172] However, in remote places where population densities are low, services cost more to deliver per capita and returns may be expected to be lower.

A history of failed project interventions has deterred governments and donors from making fresh initiatives in African drylands.[173] However, satisfactory economic rates of return (from 12 to 40 percent) have been cited for a number of projects, including soil and water conservation (Niger), farmer-managed irrigation (Mali), forest management (Tanzania), and farmer-to-farmer extension (Ethiopia).[174] Returns of over 40 percent are on record for small-scale valley bottom irrigation projects in northern Nigeria and Niger. Where financial data are not available, the impact of project interventions can be evaluated from uptake, especially in the post-project period. Such evaluations are infrequent, however. These examples draw attention to a need for better post-project monitoring, which tends to be forgotten as soon as donor interest drifts elsewhere.

Livestock investments

Animals have important value as capital in dryland farming systems, agro-pastoral and pastoral systems. Moreover such capital grows through breeding and through fattening for market. Many livestock owners do not breed but rely on markets for acquisitions. In farming systems in the Sahel, small ruminants (sheep and goats) and fowls are affordable to very poor people and women, and may be given to children as a form of saving. When contingencies call for money urgently they can be sold. In the intensive smallholder systems of the Kano Close-Settled Zone, Nigeria, livestock are stall-fed throughout the growing season, using collected residues and weeds, and their manure is systematically spread on the fields in time for the following season. Milk and animal energy are additional recurrent benefits. Free grazing is allowed in the dry season. In this way nutrients are recycled efficiently. In Machakos, Kenya, grazing takes place on private pastures and where this is not available some cattle are stall-fed throughout the year.

The economic value of livestock and of pastoralism has been illustrated in Chapter 4. Pastoralism is fundamental to the well-being of millions of drylands people. Given the low cost of inputs in rangeland systems (compared to farming), this suggests that economic returns for some livestock investments can be high.[175] Another indicator is the value of market sales of livestock products and services, which include dairy products, meat, hides/skins and wool. In farming areas, livestock also provide farm energy, transport and manure, all of which can earn income, providing a substantive return to modest livestock investments. In Kenya, a pilot project in Isiolo District implemented with a government investment of Kshs. 2.5 million resulted in earnings of Kshs. 18 million, derived from livestock marketing.[176]

Pastoralism differs from farming in two important respects. First, all available and suitable rangeland is normally in use, so no extending of pastoral territories is possible. This applies to virtually all of the worlds' drylands. The situation is aggravated by the loss of land to competing uses (farming and urbanization). Second, pastoral production

[171] (Fan *et al.*, 2000)

[172] (Hazell, 2001; Hazell *et al.*, 2002)

[173] This claim must be qualified by the recent revival of a 'Great Green Wall' project for afforestation to 'stop the Sahara', promoted by some West African heads of state with the support of the European Union.

[174] (Reij and Steeds, 2003)

[175] (Gabre-Madhin and Haggblade, 2004)

[176] (Reij and Steeds, 2003)

Transport of camels to Port Sudan, Sudan. © *Agni Boedhihartono*

systems are labour-intensive and involve the investment of human and social capital in institutions and management. For example, the WoDaaBe cattle herders of Niger practise intensive breeding based on deep local knowledge and on caring for each individual cow and her progeny (see Box 17). Grazing systems balance fodder and water availability with the capacity of animals to undertake often arduous daily journeys. It has been shown that such systems, despite the hardships imposed on their users, are more efficient in their use of natural resources than alternatives (Chapter 4). This fact is unfortunately lost by those who advocate 'modernization' in the form of large-scale ranching, a type of investment once favoured by donors who took their lead from capital-intensive systems in developed countries. Given the values of animal husbandry – and constantly buoyant meat prices driven by urban demand – livestock and pastoralism are well worthy of both private and public investment; but not at the price of ill-advised transformations.

Box 17: WoDaaBe breeding and grazing systems, Niger

The WoDaaBe have developed a cattle management system which allows them to control stress and facilitate the transmission of functional behavioural patterns within the herd (learning, feeding competence and social organisation), which they have learned by studying cattle in their environment.

The breeding population of cattle is organised into matrilineal lineages, the genealogies of which are carefully memorised. Cattle reproduction is strictly controlled. Less than 3 percent of the bulls are used for matching with all the dams, and variability is fostered. Selected sires are intensively circulated within the breeding network. Cows are rarely sired twice by the same bull. Culling of females focuses on reproductive capacity, with marketing of poorly performing animals. Selection is carried out within but not between lineages. Long-lasting lineages are sought after and protected from non-strategic marketing in case of economic pressure.

Source: (Krätli, 2008).

Investing in trees

Not often treated as capital assets by analysts, trees are a form of investment on farmers' fields (where private title is assured) and around houses, as many can generate income from NTFPs such as food (edible leaves or fruit), fodder, medicines, fibre or construction materials. While tree planting may be a sound investment, naturally regenerating trees also have asset value. Planted or regenerating seedlings must be protected from free-ranging livestock, so they do have costs.

The major use of wood in drylands is for fuel and this is followed by construction and craft timber. Because woodland is often viewed as an open access resource, fuelwood cutting has been blamed for deforestation. This is only partly true. Contrary to claims of extensive treeless 'deserts' appearing in the vicinity of fuelwood markets, the value of multi-purpose trees to their owners normally results in their protection and the displacement of commercial fuelwood demand to areas of easily accessible woodland - up to 200 km away in the hinterland of Kano, for example.[177] Markets for rights to cut dead or branch-wood encourage farmers to produce wood as a subsistence activity or as a commercial proposition. Labour inputs on protection and management are small enough to ensure a good rate of return.

The production of most NTFPs is poorly documented. Gum production (especially Gum Arabic, derived from *Acacia senegal*) is better than most as it enters international trade. In the Sudan (the world's major exporter), producers stand to gain most of its market value in profits as the gum is obtained from naturally regenerating trees on fallow fields, although transport to ports from inland locations reduces net returns. In another exporting country (Ethiopia), gum collection and sale are important to producers' livelihoods (Box 18). Another tree product of commercial importance is frankincense, also exported from Ethiopia. Although investment or input data are not available for a calculation of economic returns, the profitability of such NTFPs to rural livelihoods may be inferred.

In East, Central and West Africa, the continuing importance to household incomes, nutritional and food security of the fruit of trees such as *Adansonia digitata*, *Tamarindus indica*, *Zizyphus mauritiana*, *Sclerocarya birrea*, and *Mangifera indica* has been often noted. Dryland production of high-value fruit can stand transport costs to market and still yield a good return on investments in soil and water conservation (as in Machakos District, Kenya).[178]

The planting, protection and harvesting of multi-purpose trees is thus an important contributor to household incomes and nutrition in drylands. In place of plantations of exotic species, established by the state, and offering little to local communities, the revival of customary protection of indigenous species on private land offers more benefits to both communities and ecosystems. These benefits, and the trade-offs involved in realising them, are highly specific to species, conditions and management. Not all NTFPs are harvested from private trees, however. Trees in common access woodland are

Box 18: Investment returns from Gum Resin, Ethiopia

In Ethiopia, drylands constitute 70 percent of the landmass and livelihood options are few given the harsh environmental conditions. Therefore production and marketing of frankincense and other commercial plant gums is essential for sustainable development. Gum trees also contribute to the conservation of dryland ecosystems.

In South-East Ethiopia, the average annual cash income generated per household from sales of gum resins was estimated at USD 80. This contributes 32.6 percent of annual household subsistence costs and ranks second after livestock in the overall household budget. Crop farming only contributed 12 percent. In 1988, 663 tonnes of gum resins were exported from Ethiopia with a total value of USD 1.23million. More recently, between 1996 and 2003, Ethiopia exported 16,019 tonnes of gum resins per year, worth USD 20.5 million.

Frankincense is a resin extracted from Boswellia species. Local people have traded in it as a means of diversifying their incomes. Between 1996 and 2003, Ethiopia exported 13,299 tonnes per year of gum resins (90 percent frankincense), earning USD 18 million.

Sources: Lemenih *et al.*, 2003; Lemenih *et al.*, 2007; Tadesse *et al.*, 2007.

[177] (Cline-Cole *et al.*, 1990)

[178] (Mortimore and Tiffen, 1994)

also sought out and harvested for useful products which are sometimes sold. The value of a wide range of NTFPs, especially medicines, supports biodiversity conservation on village lands. The same applies to many herbs and grasses growing naturally in the ecosystems.

Private (commercial) investments

Current economic orthodoxy places much confidence in open markets and the role of the private sector in the development process. However, unlike the small-scale investing of private savings in dryland communities, private commercial investments are not easily attracted to drylands, and a negative stereotype is prevalent in many countries. This is due to:

- A lack of physical infrastructure (safe water, electricity, solar energy, transport network, markets, telecommunications, schools, health centres and shelter) that promote human capital and private sector development;
- Poor access to financial resources and services such as banking and credit facilities;
- Little information on exploitable ecosystem resources, costs and values;
- Insecurity;
- Distrust of local populations;
- High tariffs on long market routes;
- Insecure tenure.

But opportunities for private (commercial) investment in drylands are likely to increase with growing monetization of the local economy, international trade, rural-urban interaction and the emergence of larger middle-income social groups in cities if not rural areas. Among such opportunities are:

- Niches in market chains such as commodity bulking and out-grower schemes. Such chains have been identified as critical factors in development in the Sahel, for example;[179]
- Technology development and maintenance;
- Service provision (agricultural extension, health and possibly education provision);
- Credit provision and banking;
- Cell phone networks;
- Transport provision.

However, early expectations that the private sector would fill the gap left by public sector service providers after these were withdrawn following structural adjustment programmes in some countries have not yet been fulfilled. Higher returns are needed than those acceptable to smallholders, who tend to discount labour costs. Attention needs to be given to incentive structures if the private sector is to fill the gaps created by a retreating public sector, or to share the burden of meeting the MDGs.

However, stagnating commercial investments in rural drylands may be swiftly reversed if carbon and biofuel markets expand into the drylands on a large scale (see Chapter 6). Potential returns to investments are very uncertain under conditions of seasonal aridity, uncertain rainfall and poor soils. Whether these markets are seen as a threat to dryland populations and ecosystems or as an opportunity to attract inward investment to the development process depends on one's point of view. An important issue with private commercial investment is the difficulty of taking proper account of the social and environmental costs and benefits of new developments. Both are likely to be higher than in urban or more humid locations. The best way forward may be through partnerships between public and private interests.

Investment incentives

Investment can be enabled through appropriate policy instruments. Private investment can be stimulated by public-sector investments and policy. There are two main categories of investment incentives: direct and indirect. The first category is linked to projects, where financial gains are made through project participation. These depend on funding for projects. The second category is indirect incentives, including both *market* and *enabling* incentives.[180]

All the evidence accords a critical role to *market* incentives. Dryland peoples attach much importance to market participation. For most, the risks associated with isolation from markets (cash and food scarcity, unemployment, knowledge deprivation) now outweigh the risks of closer involvement (for example, dependence on highly priced food in times of scarcity). Closer involvement is seen to have many benefits (sales of produce; supplies of food

[179] (Bolwig *et al.*, 2009)

[180] (Knowler *et al.*, 1998)

and consumables, inputs and technologies; labour exchange; information; education-based careers; remittances; and investment funds).

Alternatively, policy can work through *enabling* incentives. These cost government very little: they are embedded in the policy framework which is configured by the political process and institutions. Among the critical institutions whose relevance is clear from experience are: land tenure, common pool resources, credit institutions, decentralized government services, and research and extension systems.[181] The scope for influencing investment depends on the architecture of a particular country's institutions, for as we have stressed, dryland countries are not all the same.

Poor dryland producers are not necessarily too poor to invest human and social capital (labour, skills, knowledge, local institutions) as well as savings into the long term. Small-scale private investments were keys to each of the landscape investment stories (above), even where public-sector investment also played a role. The context of the decisions of small investors is critical. There are opportunities and constraints facing the individual investor that reflect the enabling incentives present in the economic environment, macroeconomic policies and the risk of external shocks such as drought. Resources are allocated to meet livelihood objectives (which include other elements besides agriculture), taking account of the costs and expected benefits (e.g. to present or future income, leisure, inheritance). Many considerations, in addition to financial returns, have a bearing on these decisions. Among them are consumption requirements, social obligations and off-farm income opportunities. Many constraints, however, impede investment, including risk, lack of funds, soil infertility and ignorance of markets or off-farm alternatives. Thus, natural resources are embedded in a livelihood investment framework.

Policies to promote dryland investment face a major challenge in the form of high perceived levels of risk. The biggest source of risk is a variable climate, which may directly cause losses of livestock or crops from droughts or floods, with ramifications throughout the local economy.

Risk management in drylands

If we are to understand dryland development in terms of an ecosystem that cannot be predicted to return to equilibrium after a disturbance, but may change from one state to another, then resilience is a requirement in the human system that is coupled to it in complex ways.[182] We have observed (Chapter 2) that this property is manifested in adaptive strategies that are mobilised particularly in droughts but are equally important in managing the impact of other adversities originating in either the natural or the human systems (such as floods, pests, sickness and death in the labour force, conflict, sudden market failure, or unanticipated policy changes).[183] Local drought occurs almost every year in China. In Mongolia, the biggest source of risk is prolonged heavy snowfall (*zug* events) which can block access to grass and cause starvation.[184] Drought (and other) risks are therefore linked explicitly with development, but dryland communities usually lack a strong political voice.[185]

Pastoralists in African countries have customary loaning and insurance arrangements, and working institutions for redress and debt collection. These assist individual households to survive in bad times and to rebuild after losses. Such institutions may be enhanced through policies to broaden micro-credit and investment opportunities, and to enable access to financial services such as banking and insurance.

However, market failures can work in the opposite direction. High transaction costs, resulting from poor information flow to producers, a lack of competition in the supply of goods and services, and an inability to choose the best time of sale, disadvantage pastoralists when they want to convert their livestock wealth in times of climatic stress.[186]

Small-scale farmers have evolved a range of coping strategies to help them adapt to variable rainfall. In the Sahel, these include on-farm responses (changing crops or crop varieties, water harvesting practices, and the production of surpluses whenever possible for storage) and off-farm strategies (harvesting more wild products from the ecosystem, including famine foods, sale of manufactures such as mats or ropes, seeking work as hired labour, and migration to other places for trading

[181] The role of institutions in drylands is take up in Chapter 7

[182] (Liu *et al.*, 2007)

[183] (UNDP-DDC, 2006)

[184] (Chuluun, 2008)

[185] (Wily, 2006)

[186] (WISP, 2007b)

or employment).[187] In Central and West Asia and North Africa, strategic on-farm priorities have been identified as: changing cropping systems and patterns, switching from cereal-based systems to cereal–legume mixtures, introduction of drought and heat resistant varieties, use of more water-efficient irrigation practices, adopting existing or new water harvesting technologies.[188] Small-scale farmers in the high *altiplano* of Bolivia, Peru and Chile developed effective systems of adaptation to extreme climatic conditions based on crop diversity and ecological zoning.[189] Other opportunities for diversification in drylands that have been suggested are: adding value to livestock products through rural based processing industries and genetic improvement, mining, fishing, eco-tourism and cottage industries, apiculture, and bio-fuels.

Crop insurance, as a strategy to reduce the risk of crop failure from adverse weather, and avoid prematurely committing farm resources, has not yet been extensively tested in drylands. Contracts with farmers are based on the mean local rainfall.[190] If rainfall is below this value at a critical time of the cropping season, then all who have purchased insurance receive compensation. Such schemes have been piloted in Malawi (from 2005), India and Ethiopia, but it is too early to conclude on their efficacy, especially for the poorest farmers. In Mongolia, index-linked insurance schemes are in operation for the livestock sector.[191] In the Horn of Africa, however, emergency livestock marketing interventions, popular with NGOs, enter a complex world of existing drought responses. As always, livelihood strategies already in place should provide a platform for such interventions.

All insurance schemes, whether customary and local or modern and more extensive, face the constraint of co-variance in climatic events across wide regions. If all livestock producers, or all farmers, - or worse still, both - suffer losses simultaneously, compensation schemes may collapse locally or become entirely dependent on external assistance. This, in effect, is what international food aid already does - though often too late to avoid high costs from the inflation of commodity prices.

Another option for the public sector is better (more relevant) seasonal rainfall forecasting to minimise losses from committing resources before rainfall outcomes are known. Progress has been made in the technical development of forecasting in Africa, and in providing simple messages that farmers can use in making decisions about their inputs.[192] However both the demand for and benefits of weather forecasting, for both farmers and pastoralists, have not been adequately demonstrated on a wide scale.

Conclusion

There is a fundamental tension between high levels of environmental risk in drylands and a need for acceptable returns from investment. How can the quality of resilience be imparted to an investment system? If assets are protected through droughts (or other crises), investments can be cumulative, and development becomes a reality. If not, asset divestment in food emergencies will frustrate growth. Insurance in some form is key to an upward cycle of cumulative investment. While indigenous capacities for risk management have been under-valued by policy makers in the past, they are not sufficient on their own to underwrite development in uncertain times. However the challenge to devise effective management led by appropriate public policy is not insuperable, as recent international consultations show.[193] 'Political will' is essential as many drylands have suffered from investment neglect. Risk management, investing in people, their capacities and their institutions, and creating an enabling environment are the most appropriate goals for policy makers.[194]

In searching for a new policy paradigm for drylands, markets will play a fundamental role. This theme is taken up in Chapter 6.

[187] (Mortimore and Adams, 1999)

[188] (Thomas, 2008)

[189] (Bolivia, 2006)

[190] (Hazell *et al.*, 2002)

[191] (UNDP-DDC, 2006)

[192] (Cooper *et al.*, 2008)

[193] (UNDP-DDC, 2006; UNDP-DDC, 2008)

[194] (Hazell, 2001)

Local products, Uzbekistan. © *Daniel Kreuzberg*

CHAPTER 6
Linking drylands with markets

Drylands in North and Tropical Africa and Asia have deep historical ties with markets, cities, and distant places. Central Asia and the Sahara were criss-crossed by trading routes linking East with West, and tropical environments with temperate. High value commodities, new knowledge, slaves and conquering armies traversed them; silk and printing technology travelled from China to Europe, and in West African history, the savannas lent themselves to the passage of horse-borne empire builders. Nomadic populations of the deserts and steppes played intermediary roles in these exchanges, and Arabic culture was created in and exported from drylands to the humid biomes of the Old World.

So it is a myth that drylands - in recorded history, if not earlier - always acted as barriers to economic, social or political intercourse, and only had a history of remoteness and isolation.[195] It is important to bear this in mind when confronting the relative marginalisation of many dryland peoples in the contemporary world. Markets have enjoyed a central place for the past two decades in the ideology of development (the 'Washington Consensus').[196] At the local level they play an important role in the everyday life of dryland communities. Participation in markets has increased, but many would say that it is failing to bring development on the scale expected. We shall offer a brief analysis of the ways in which this participation is changing, and the opportunities that sound investment and policy can exploit.

The demise of colonial export agriculture

The promotion of cotton, groundnuts, and tobacco were central to agricultural policy in many African drylands from the beginning of colonial rule. A symbiosis between growing demand for fibres, vegetable oils and stimulants in the industrial countries on the one hand and, on the other, land-surplus economies with a need for monetary rewards (with which to pay taxes and to finance rising consumer expectations), created new opportunities. Governments installed transport and port infrastructure, financed the introduction of new varieties, set up agricultural extension systems, and facilitated processing plants (cotton gins, oil mills, tobacco factories). This system peaked in the 1960s (the decade of independence). Thereafter, cotton exports became less remunerative as world prices stagnated and the demand from local textile mills escalated; groundnut exports were decimated by rosette disease, drought and falling world prices; and tobacco production was also diverted to local markets even as northern producers came under pressure from anti-smoking lobbies. Under the impact of falling world prices, the state-driven systems of marketing, processing, credit provision and technical support failed (most dramatically in Senegal).[197]

This model helps to explain the diminishing share of African countries in world trade, though it does not apply equally to all export crops nor to all drylands. It has been reinforced by the reluctance of the USA and Europe to terminate agricultural subsidies and market barriers. The European Union's preferential trade agreements, originating in bilateral colonial relations, have been dismantled in favour of its current policy to impose 'free trade' agreements in the name of development. American subsidies to its cotton farmers are believed to exceed the national income of Burkina Faso, a dryland country that still depends on cotton exports. Dryland futures will be insecure if built on low value export agriculture.

On the other hand, colonial export agriculture was based on rapid expansion of the cultivated areas, as subsistence farmers were reluctant to substitute export for food crops. Since land was abundant and free, woodland was cleared and soil nutrients 'mined' and exported. Concern about the longer term implications of this were met with subsidized inorganic fertilizers.[198] The eventual exhaustion of

[195] The Australian drylands perhaps come uniquely close to the stereotype, but on account of oceanic, not desertic barriers.

[196] (World Bank, 2000)

[197] (Faye *et al.* 2001; Faye, 2008; Meagher, 1997)

[198] (Franke and Chasin, 1980; Watts, 1983)

supply in unclaimed land has forced sustainable soil management to the top of the agenda (Chapter 2), although the political economy of nutrient cycling attracts surprisingly little attention. Markets play a major role in this unfolding narrative.

New commodities for growing and urbanizing populations

Concurrently with the decline of colonial trade relations, population growth was approaching a 30-year doubling time in some African countries, and urbanization was accelerating rapidly throughout the world's drylands, with accompanying currents of short- and long-term migration. And while concern was (and is) being expressed about the growth of urban poverty and the deterioration of urban environmental quality, less attention was given to the implications of this demographic transformation for commodity markets. Urban dwellers buy rather than produce food and other rural products, and their low average incomes notwithstanding, they generate significant growth in aggregate demand. In an absence of state provision, this demand drives expanding informal market systems, privately-owned public transport, and extending urban hinterlands. For example, in West Africa's drylands between 1960 and 1990, market growth correlated with population density and output per ha and per rural inhabitant, indicating a strong coherence between these variables (Box 19).

From 1960 to 1990, at annual population growth rates averaging 2.4-2.6 percent, and rates of urbanization increasing from 8-10 percent to 30-32 percent, the settled Sahelian zone of West Africa saw the number of towns with more than 100,000 inhabitants increase from five to 25. An indicator of 'market attractiveness' (based on size of market, distance from market, costs, competition and supply determinants) shows a dramatic increase in the areas linked 'strongly' or 'moderately' to markets. This trend is expected to increase, reaching 90 percent of the settled Sahelian zone by 2020. Rural population density correlates strongly with market attractiveness, and so does output per ha and per rural inhabitant.[199]

Box 19. Urban provisioning in Kano, Nigeria

Food commodity markets were followed from 1960-1999 in Kano, one of the major cities of the region. Despite the vicissitudes of policy changes and climatic variability, the markets demonstrated flexibility and a capacity to supply growing numbers of consumers with staple cereal grains and livestock products, at prices that tended downwards in real terms. Rural-urban interaction was not new. Since the early nineteenth century or before, food commodity producers had supplied the city from its immediate hinterland using intensive organic farming methods supported by livestock, which was considerably helped by exports of manure from the city to the countryside. By the middle of the twentieth century, however, the sources had moved away to distances of 100 km or more as local farmers sought to feed their growing families from the soil. Kano is now a hub in the regional grain trade.

Sources: Ariyo *et al.*, 2001; Mustapha and Meagher, 2000.

African regional economic groupings such as ECOWAS and SADC[200] encourage interstate trade in food commodities. Comparative advantage can be reaped by producers. This is highly relevant in a context of rapid but localised urban expansion. From an ecosystem management perspective, the movement of plant nutrients remains a critical issue but at a reduced spatial scale.

Niche markets – a return to exports?

Dryland ecosystems offer a range of high value natural products whose development may be facilitated by globalization. Access to niche markets for eco-friendly products, using trademarks, can earn premium prices. There is an increasing demand for innovative, unique or speciality products aimed at these markets, worth globally about USD 65 billion per annum, which features strongly in some dryland country economies. A significant proportion of this demand is for medicinal and cosmetic products. Global trade (including industrial production) in aloe, a skincare product traditionally used in and originally sourced from drylands, is valued at about USD 80 million.

[199] (Cour and Snrech, 1998)

[200] Economic Community of West African States; Southern Africa Development Community

Other examples are Devil's Claw, a medicinal plant used for arthritis from Namibia and Botswana, valued at over USD 31 million in European Union markets; Gum Arabic (see Chapter 4); and crocodile skins, an emerging opportunity for Zambia, generating up to USD 2 million after an initial investment in 2005. But whatever the economic benefits from their exploitation, these and other potentially marketable species are listed by the Convention on International Trade in Endangered Species (CITES). Monitoring is essential to ensure that all trade, at least at an international level, is sustainable.

In southern Africa, the current natural product trade is estimated at USD 12 million per annum, with potential to grow to USD 3.5 billion, half the value of current agricultural exports from the Southern African Development Community (SADC) region.[201] A growing sector of commercial natural products employs up to nine million casual, largely female, workers (Box 20).

Niche markets for ecosystem products do indeed allow local people to make money. However, the risk of creating scarcities of valued products through exploiting common access resources unsustainably is a real one. It suggests that private interest may not coincide with the common good where financial gain is to be made. A growing market thus raises institutional questions that will receive attention in Chapter 7.

Box 20. Mongongo production, marketing and impact in Zambia

Mongongo fruits are widely known and distributed in southern Africa as a food source. Kalahari Natural Oils (KNO) makes products from Mongongo oil to treat dry skin and hair, supplied by large producer groups in western Zambia. KNO is a member of PhytoTrade Africa, partners with the International Union for the Conservation of Nature (IUCN). As a result of this support, new and expanded groups have been organised and trained to supply Mongongo in an effective, environmentally sustainable and profitable manner. A factory began operations in 2006 near Lusaka and employs five personnel producing 50 kg oil a day from 200 kg of kernels. Hair and body gel products are marketed in Zambia in 100 ml packs priced at USD 1.50. Nearly 90 percent of producers are women.

Mrs Berthe Monde harvested in one four-month season a crop of 450 kg of kernels, and sold them to the KNO for USD 450. The income was used for school fees and care of seven children, including three orphans. She learned the skills of cracking and storing Mongongo at the age of seven. She also uses the pulp to make beer or porridge for use in times of drought. Although most women knew how to use it for a cooking oil and relish, it has only recently become known as a source of substantial income. When maize (the main food crop) fails, such income supports her family through the ensuing lean period. Her profits have been invested in two plough oxen, two goats and in education (through the parent-teachers' association). Her long-term aim is to build a better house.

Source: K. Faccer, PhytoTrade Africa/ IUCN Natural Futures Programme.

Livestock markets

Given the importance of pastoral production systems in the drylands, and the role of livestock producers in meeting a globally rising demand for meat, milk and other livestock products (the 'livestock revolution'), it is necessary to take special account of the modes of engagement with markets that are found in different dryland regions. Rather than producing animals to supply markets, as in commercial ranching systems (where the most financially efficient off-take levels are sought), pastoralists seek to use the market as an instrument in achieving their wider livelihood objectives, in particular building up and maintaining high quality herds in an environment that can be dangerously variable. Thus among the WoDaaBe of Niger, buying and selling animals is intimately embedded in selection and breeding strategies.[202] In the Ferlo of Senegal, the Fulani embody their awareness of risk in their engagement with markets (Box 21).

Herders may suffer disadvantages in marketing, as in Tibet, where strongly seasonal patterns and difficulties of accessing markets on favourable terms of trade work against their interest in increasing their level of participation (Box 22). Nevertheless, increasing participation is both necessary and inevitable.

[201] (Bennett, 2006)

[202] (Krätli, 2008)

Milk of Eritrean herder in Sudan for local market. © *Agni Boedhihartono*

Box 21. Risk and markets in the Ferlo, Senegal

The pastoralists at Tatki (Chapter 4, Box 9) take account of all of their needs and levels of risk (subject to the information available to them), and the condition of the rangelands, in choosing the best strategy to meet their pre-determined livelihood objectives. The same approach is applied to markets. Rather than maximising profits, they take a 'self-bounded' attitude to sales determined by their needs, especially for cereals, and known risks. They may aim for profit, however, in situations where they receive information and favourable market signals. This is seen in the ram market before Aïd El Kébir.

Source: Wane *et al.,* 2008.

In Northern Nigeria, the marketing decisions taken by owners of livestock have tended to increase, in aggregate, the percentage off-take from herds over the longer term. In the range 9-11 percent/ year, this level probably indicates a maximum, given the risks inherent in drylands. In Maasai land in Kenya, agriculture has advanced but livestock numbers have remained fairly static, and livestock marketing rates have risen greatly.[203] Any study of market behaviour gives the lie to the myth that in Africa, animals are kept only for status, and the more the better. A correct understanding sees animals as part of broad livelihood strategies in which subsistence continues to be a primary concern of families.

Rising prices for cashmere wool have encouraged an increase in goat and yak herding in Mongolia since the transition to a capitalist economy from 1990, and in Tibet.[204] The huge Chinese market is critical. Vicuña fibre from the high tropical Andes finds a valuable export market, especially in Italy, providing an important opportunity for rearing these animals as a part of the livelihoods of local communities.[205]

[203] (Norton-Griffiths, 2007)

[204] (Chuluun, 2008; Nori *et al.*, 2008)

[205] Robert Hofstede, *pers.comm.*

Box 22. Marketing livestock products on the Tibetan Plateau

Despite increasing needs for cash and growing market integration since the advent of an open market system in the 1990s, herders still orientate their management to subsistence. The level of market participation depends on available surpluses and the accessibility of markets. There is a strongly seasonal pattern in marketing: in summer, herders sell cashmere, hair and wool to buy domestic items and food if needed. In autumn they sell animals or meat, dairy, skins and dung, in order to buy imported food for the winter. Health needs, new taxes and technical innovations (such as solar panels) generate an increasing need for cash. But problems of access and seasonality tend to turn the terms of trade against the herders, and they are vulnerable to external forces such as price fluctuations, poor transport networks and inadequate information.

Source: Nori, 2004.

In addition to the mobile pastoralists, farmers (or agro-pastoralists) in drylands keep large numbers of livestock, especially sheep or goats, and buy and sell according to circumstances (e.g., buying after harvest and selling at price peaks before religious festivals). A capability for such flexible responses to variability is necessary if livestock are to help in building livelihoods.

Dryland ecosystems tend to behave in an unpredictable manner (Chapter 2). Livestock depend mainly on natural biomass and are vulnerable to unpredictable fodder shortages. This has led many to conclude that their capacity to damage the ecosystem through 'overgrazing' is limited, while the risk of loss for their owners is high. As they cannot prevent such losses, their strategy is opportunistic, increasing the numbers of animals in the good years so as to maximise the chance of some surviving in the bad. Such strategies must form the basis of new developmental initiatives.[206]

As a consequence, livestock owners have a more erratic relationship with markets. They must sell animals at unfavourable prices when food prices peak, and buy animals after the crisis when their price has risen. Whereas crop commodity markets can promote unsustainable use of the soil, livestock product markets depend on rather than determine the condition of the ecosystem.

Value chains in drylands

For every marketed commodity there is a value chain (*filière*) linking producers with end-users through intermediaries. At each stage value is added, so that the interests of producers are served by efficiency gains through competition or regulation. Conversely, market failures (such as monopolies, illegal rent-seeking, excessive taxation, or withholding fair prices from women) inflate end-user prices or deflate producer prices. Along these chains, therefore, are found the opportunities to regulate or intervene in support of poor producers of crops, livestock, natural products or others.[207] For example, emergency relief interventions can be designed in terms of a model of the value chain seen as embedded between environmental drivers on the one hand (such as weather, taxation) and internal services on the other (such as transport, credit).[208] The form such a model takes is specific to a particular time and place.

Fair trade and organic certification initiatives are intended to increase producers' gains on internationally traded products. In the Sahel, they are rare but increasing, as higher value products find markets.[209] Value chains are changing rapidly, especially at the international level, and are a key entry point for development.[210] A South American example is provided in Box 23.

Value chains that link producers with urban consumers within the same country or region are growing in importance with urbanization and cross-border trade (for example, see Box 19). There are positive linkages between these value chains and investments in intensive farming, including soil and water conservation. African examples include Machakos in Kenya,[211] and the impact of urbanization has gone even further in India, where specialisation in food commodities responds to a range of well defined market niches. These complex linkages offer an alternative vision to that which sees market pressures only as a threat to ecological sustainability.

---

[206] (Sandford, 1994)

[207] (WRI *et al.*, 2005)

[208] (Jaspars, 2009)

[209] (Bolwig *et al.*, 2009)

[210] (Vermeulen *et al.*, 2008)

[211] (Tiffen and Mortimore, 2002)

Box 23. Vicuña fibre from the Peruvian Puna to the European fashion market

Vicuña is a cameloid species, native to the South American high mountain drylands (*puna, altiplano*). Its fibre is one of the highest valued in the world, especially in European markets. Commercial farm management of Vicuña is practically impossible, and fibre was obtained by hunting the wild animals, driving the species almost to extinction. By putting it high on the CITES list and introducing protection measures in Peru, Bolivia, Chile and Argentina, the species was saved and the natural population recovered.

A state-controlled management programme was recently set up in the Reserva Nacional de Salinas y Aguada Blanca in southern Peru. In collaboration with local communities, a way of managing wild populations of Vicuña was developed, based on 'rodeos' without the need to kill the animals. Italian buyers, organised through the International Vicuña Consortium, maintained quality control and bought the product at USD 300/kg, and provided USD 700,000 for investment by the communities. A controlled market chain was set up providing a new and sustained income for the local communities within the overall conservation of the Park.

Source: Peru, 2008.

Biofuel production

New potentials are opening up for producing biofuels in drylands. These are being driven by targets set in the USA, Brazil and Europe for achieving a minimum biological fraction in petrol and diesel fuels. In Brazil, large areas of woodland have been cleared for biofuel production. African drylands are also perceived as having abundant unused land. In particular, *Jatropha curcas* is perceived as an option under low rainfall, even in the pastoral zone where other crops are not economic. In fuel-importing countries, especially those where major centres of economic activity and urbanization are a long way from the ports, biofuels appear to offer a means of reducing import dependency. Such a country is Kenya (Box 24). The possibility of even a partial relaxation of the economic stranglehold exerted by fuel bills is attractive to economic planners. But in Kenya, most of the rural population live on about 15 percent of the territory, and the remaining drylands are used by pastoralists or are hyper-arid. Pastoralists tend to be invisible to planners and optimistic assumptions of available cultivable land will be disputed.

In Brazil and the USA, subsidised biofuels already compete directly with food crops, with a negative impact on global food security. In South America, the feasibility of *J. curcas* is being studied in Ecuador and Peru. So is soya in the *pampas* of Argentina. Other possible impacts are pollution from increased use of fertilizers and pesticides, and risk of displacement of livestock production into the forests.

In Africa, the idea of drylands as 'wastelands' has been pervasive. But the amount of cultivable but unused land still available is small, after more than a century of agricultural expansion, and most uncultivated areas are subject to customary grazing rights. Pastoral interests are already under pressure from agricultural expansion and urbanization. If prices are favourable, biofuel production will displace food production on existing farms. The producers who will be attracted to biofuel production are less likely to be the poor, who value their subsistence production most highly, but rather the better off, who can risk capital, buy land and afford to buy food. An additional challenge is regulating new and novel markets, especially where outgrowers are signing unfamiliar or ill-defined

Box 24. Biofuel in Kenya

A proposal targeted at economic planners claims that if Kenya were to offset 10 percent of petrol imports and 2 percent of diesel imports with locally produced biofuels by 2013, it could keep USD 71 million from flowing overseas, given a petroleum price of USD 90 per barrel. Within five years, Kenya could achieve these targets, plus providing surplus production for stationary power and exports. Depending on land availability, yield and economics, 15,000 ha of 'new' land would be dedicated to sugarcane (16.5 percent of suitable land that is not currently being used for food or cash crops), and 'a portion' of current sugarcane production would be diverted from food production. Sweet sorghum would be planted on 24,700 ha (1 percent of suitable 'new' land). For biodiesel, a mix of castor, coconut, croton, jatropha, rapeseed and sunflower would require about 50,000 ha, some of which is already planted. It is claimed that biofuels could revitalize rural areas, like Nyanza and Western Provinces, and provide an engine of growth (see text for a critique).

Source: GTZ, 2008.

contracts with intermediaries, such as in Zambia.[212] Finally, new pressure on uncultivated land will threaten biodiversity. In Kenya, where most wildlife is now concentrated into reserves, it is diminishing outside the reserves as agriculture expands.[213] These and other questions show that unless supported urgently by appropriate policies, a 'biofuel revolution' in drylands may be a mixed blessing for both livelihoods and ecosystems.

Carbon markets

Carbon markets in Africa under the Clean Development Mechanism (CDM) are underdeveloped because of problems of eligibility, investment barriers (which are not peculiar to the CDM), and low rates of return.[214] The voluntary market is new and small. African drylands are not alone in suffering from under-investment (see Chapter 5), but Africa's share in carbon markets is growing. Nevertheless, a question remains as to whether drylands can compete in carbon markets against biomes with higher per hectare potentials.

The world's rangelands also suffer from ineligibility for carbon schemes.[215] Yet sequestration potentials are significant, especially in soil carbon. As they vary in climate, soil and vegetation, rangelands differ, but management practices can increase capture (protecting trees, adding organic matter, reducing soil respiration or erosion). Households with different capital and resource endowments differ in potential, and incentives and subsidies may be necessary. Legal rights over rangelands, and clear developmental benefits, are preconditions. Attractive scenarios are available (Box 25) but scepticism may be justified with regard to the quantitative estimates.

Agroforestry can capture up to 40 percent as much carbon as primary forest.[216] Poor farmers need to be given the same incentives to sell their carbon as a commodity as they enjoy for other products. Allowing them to do so could generate as much as USD 10 billion, and agroforestry is estimated by the IPCC to have the potential to remove 50 billion tonnes of carbon from the atmosphere. Remote sensing technology can be used to monitor implementation, and the carbon markets will provide incentives to stop deforestation, which is less profitable in the long run.[217]

Carbon markets are a form of Payment for Environmental Services (PES), which recognise external interests in dryland ecosystem management. PES has long been used to provide incentives for land use decisions, for example, in upper river catchments that benefit downstream users. The capacity of dryland vegetation to capture carbon is less than in some biomes (e.g., tropical rain forest), but the extent of dry woodland and grassland in the tropics is vast. Access to global carbon markets is being promoted as a 'triple-win' strategy, making a significant contribution to climate change mitigation while bringing a new source of income to dryland peoples, and providing incentives for sustainable ecosystem management. Working schemes are in operation, such as Scolal Te in Mexico.[218] However, equity issues have been neglected in debates that underestimate the complexity of local impacts.

Labour markets and financial flows

Collaborative or communal work on farms or with livestock, based on principles of reciprocity, is a customary practice in many dryland systems. It enables a flexible response to urgent, time-constrained, large-scale or emergency tasks and

Box 25. Carbon capture scenario for the Tibetan Plateau

Alpine meadow covers more than 58 Mha on the Tibetan Plateau, and contains between 25-53 tC/ha, more than 90 percent of which is in soils. 18-year grazing studies show that continuous heavy grazing leads to a halving of soil C stocks. Official figures suggest these grasslands are overstocked by 30-40 percent. Carbon finance could play a role in providing herders with an incentive to reduce stocking rates. A policy of contracting grassland to households has been implemented in most areas. The average household has clear user rights to more than 110 ha of grassland. Average incomes are below US$1 per day. If reductions in stocking rates could increase soil C sequestration by just 0.5 tC/ha/year, then at current carbon prices a herder household might be able to receive payments of over $7000 per year, more than twice their current annual income, while also preventing the loss of important ecosystem services in this critical region.

Source: Wilkes, 2008.

---

[212] (Muyakwa, 2009)

[213] (ILRI, 2007)

[214] (IISD, 2008)

[215] (Tennigkeit and Wilkes, 2008)

[216] (ICRAF, 2008)

[217] These estimates include non-dryland areas.

[218] (WRI, 2005: p. 119-122)

represents a form of adaptation to a rapidly changing and uncertain environment. With the monetization of the rural economy, labour markets developed. In some West African drylands, analysts see a social distinction between richer and poorer farmers reflected as 'labour hiring' versus 'labour selling'.[219] Where there was a decline in communal arrangements, people would both hire others' and sell their own labour at different times in the year. In the past, some pastoralists engaged labour by maintaining a lower caste of servants. But today, money changes hands more often, though the persistence of labour-sharing institutions tends to be greater among pastoralists than among farmers.

Diversification of livelihoods is a strongly established and increasingly popular strategy for managing risk and increasing monetary incomes. Employment in cities, or areas of commercial agriculture, provides incomes that may be taken home or remitted to support consumption or investment by the family. Because income diversification usually calls for travel over long distances, the family must agree on the distribution of responsibilities. Because the drylands have long dry seasons, migrants may circulate between home farm and workplace on a seasonal basis. For livestock producers, on the other hand, labour for lifting water and conducting animals between wells and grazing fields is at a premium during the dry season.

Labour markets, the migrations associated with them, the financial flows from urbanised or wealthier regions to the rural drylands, and the adjustment of customary labour sharing institutions represent key developmental opportunities for households and rural communities. Yet rather than facilitate such spontaneous adaptations, the state has often obstructed free movements, deplored the arrival of migrants in urban areas, ignored their contributions to urban economies and markets, and even forcibly driven them home, especially if they came from across an international frontier. New policy thinking could reverse this failure by adopting a new approach to the necessary symbiosis between drylands and core economic regions. Certainly the permanent loss of labour may threaten the sustainability of rural production systems. However, the investment of remittances can increase their productivity, underwrite social claims to land, and (as noted above) motivate sustainable ecosystem management.[220]

Land and other natural resource markets

Concurrently with the monetization of labour sharing and the diversification of incomes has come the emergence of land markets, and to a lesser extent of markets in other natural resources such as trees, woodland, and water. Long ago, valuable 'point' or 'patch' resources, such as land, water and trees in oases or wetlands, were exchanged or allocated as political favours, with potential for disputes or actual conflicts. A more recent tendency is for rights to natural resources including land to be arbitrated by markets, facilitating both accumulation by more wealthy individuals and impoverishment of the poorest.

Given the observed pressures of rising demand, dryland natural resources will be driven increasingly by market values in future. Development policy needs to obtain a better understanding of the stakeholders and their interests and should aim for a regulatory role for the state that is even-handed and transparent. Greater empowerment of dryland resource users is a necessary first step in the inevitable negotiations.

Land and natural resource markets both threaten customary users, reliant on common rights, and provide private security for investment. Individual private and exclusive rights are not, however the only or the best solution in every situation. Collective institutions still have a future (see Chapter 7).

Input, service and knowledge markets

With the withdrawal of many states from production and service delivery in the agricultural sector, which characterised structural adjustment programmes in many dryland countries from the early 1980s, a gap appeared which has not yet been filled by the private sector. Drylands lose out because they include extensive sparsely inhabited areas, under-supplied with public investment, infrastructure, service providers and access to new knowledge. Given low levels of bio-productivity and high levels of risk, incoming private investment may likely prefer cities, irrigation or other high value opportunities. Nevertheless, success stories have been reported from India having impact on large populations (Box 26).

---

[219] (Hill, 1972)

[220] (Mortimore and Tiffen, 1994)

Box 26. e-Choupal in India

ITC, one of India's leading private companies with interests in agribusiness and packaged foods, designed the e-Choupal system to address inefficiencies in grain purchasing in the government-mandated marketplaces known as *mandis*, in several states.

"Traders, who act as purchasing agents for buyers, control market information and are well-positioned to exploit both farmers and buyers. Farmers have only an approximate idea of price trends and have to accept the price offered them at auctions on the day they bring their grain to market. The approach of ITC has been to place computers with Internet access in farming villages, carefully selecting a respected local farmer as its host. Each e-Choupal ['gathering place'] is located so that it can serve about 600 farmers. Farmers can use the computer to access daily closing prices, as well as to track global price trends or find information about new farming techniques [or] to order seeds, fertilizer, and consumer goods from ITC or its partners, at prices lower than those available from village traders. At harvest time, ITC offers to buy crops directly from any farmer at the previous day's market closing price; if the farmer accepts, he transports his crop to an ITC processing centre, where the crop is weighed electronically and assessed for quality. The farmer is then paid for the crop and given a transport fee.

Compared to the *mandi* system, farmers benefit from more accurate weighing, faster processing time, prompt payment, and access to a wide range of price and market information. Farmers selling directly to ITC. . . typically receive a price about US$6 per tonne higher for their crops, as well as lower prices for inputs and other goods, and a sense of empowerment. [In 2004], e-Choupal services reached more than 3.5 million farmers in over 30,000 villages".

Source: Annamalai and Rao, 2003, cited in WRI *et al.*, 2005: pp. 102-3.

Conclusion: markets can work either way

Economic diversification is driving poor people into greater participation in markets - and not only in drylands.[221] Strategic investments of time (labour) as well as savings in alternative livelihood options have provided more than an escape from the potentially dire consequences of extreme events, but also a development pathway that deserves far more recognition from governments and donors. Producers of natural products in Zambia, for example, have found ways to shift investment strategies while minimizing risk.[222] There is evidence that 'populations are not passive victims of their environment, but have excellent coping capacities, are innovative and extremely responsive to economic signals and activities'.[223]

Markets can either distort or support sustainable ecosystem management. For example, the dum palm is over-exploited in northern Nigeria as it is regarded as a common access resource for timber and fibre. On the other side of the border, in Niger, dum woodland was protected by forestry legislation until recently.[224] In Bolivia, attractive prices for natural sisal bags encouraged women to abandon subsistence farming and exploit the plant to destruction; and in Africa, bushmeat hunting for a niche market has resulted in a reduced population of primates.[225] Sustainability cannot be guaranteed from market-based development unless the trade-offs are anticipated. This indicates a need for governance and appropriate, effective institutions. In the next chapter we turn to these.

[221] (Bryceson, 2002)

[222] (IUCN, 2007)

[223] (Dobie, 2001)

[224] (Mortimore, 1989)

[225] (WRI *et al.*, 2005: p. 105)

Granaries are an essential feature of the production system which relies on storage of at least one year's food grain, Diourbel region, Senegal. © IUCN Photo Library/*Michael Mortimore*

CHAPTER 7
Rights, reform, risk and resilience

Ecosystems are 'the wealth of the poor'.[226] But this wealth must be safeguarded by legislation and institutions to regulate access to their services, to secure their maintenance, and to apportion benefits. Under the accelerated global dynamics of colonialism, socialism, and resurgent capitalism, the twentieth century especially saw customary institutions come under stress in many drylands. Furthermore, the number of people at risk from environmental variability increased. What legislative or institutional response is needed to release the full potential of dryland people to secure their rights to ecosystem services, adapt to risk and change, and build more resilient livelihoods?

The issues

Ecosystems provide the following services to dryland communities:

- Biodiversity;
- Soil organic chemical and physical properties;
- Water holding capacity, run-off and infiltration;
- Micro-climate regulation;
- plant communities for food, fodder and other uses;
- carbon sequestration and storage;
- wild fauna; and
- spiritual and cultural value for indigenous people and (in some places) tourists.

These services support three major livelihood or land use systems:

- farming, often with subsidiary animal herding;
- animal herding (pastoralism), sometimes with subsidiary farming (agro-pastoralism); and
- wild harvesting (hunting, fishing, gathering), practised in combination with others or (more rarely) exclusively.

The legislative and institutional framework that supports these land use systems should therefore accommodate their basic rationales, as customary institutions do. Unfortunately, this has often not been so. Central to these rationales is both the 'normal variability' of a strongly seasonal ecosystem - wet and dry, abundance and scarcity of biomass - and the 'expected unpredictability' of droughts (and, occasionally, floods).

Pastoralism depends on mobility between scattered and variable pasture resources, wetlands, salt licks and markets. In all of the world's drylands, grazing rights have been based on community membership and customary recognition. In East and West Africa, under controlled access tenure systems, access to water, pasture or salt is managed according to their scarcity and productivity. High value resources such as dry season water or tree fodder are often viewed as clan or individual property within a communal system, but wet season pastures and surface water are viewed as common property with fewer conditions attached to their use. Such nesting of tenure is an efficient way of controlling access to critical resources.[227] In times of stress (such as droughts), normal migratory patterns must be adapted urgently to minimise the loss of animals. Nevertheless, large losses occur from time to time, and herd rebuilding takes several years, particularly if households lose core reproductive animals. But the flexibility of customary systems - which challenge contemporary notions of property, land use and nationhood - has often run into disfavour with governments, which have tried to reduce or eliminate pastoral mobility and induce 'better' or 'modern' systems.

Small-scale ***farming*** in drylands begins with forest clearance and extensive, shifting cultivation that reflects unimpeded access to land, and a scarcity of labour relative to land. Farmers have thus been blamed for deforestation, along with their counterparts in the humid forests. This system, in which soil fertility after cropping is restored by naturally regenerating woodland, is threatened by increasing demand for land as population grows and commodity markets develop. Fallows shorten and soil nutrients decline. Farmers are thus blamed for soil degradation, and in many places, erosion. States have attempted to 'rationalise' smallholder

---

[226] (WRI *et al.*, 2005)

[227] (Lane, 1998; PAGRNAT, 2002; Turner, 1999b)

farming by affirming the sovereignty of the state over customary land rights, by reforming land tenure from its basis in communal or family rights to individual and private rights, and by controlling input and output markets and credit.

Where smallholder rights were weakly protected and their numbers small, or European settlement was to be accommodated, states allocated large-scale commercial farms and ranches (or *latifundia* in South America). These were supported by land grants under statutory tenure, which often conflicted with indigenous perceptions of ownership. They froze extensive forms of land use over large areas, while pressures built up outside (for example, in the reserves of eastern and southern Africa), creating a strongly dualistic institutional framework.

Wild harvesting is important for many dryland communities, and particularly for women and poor people, but it is threatened by land appropriation for public or private livestock projects, and may be impeded by parks or reservations. Even in the absence of such interventions, wild harvesting depends on common resources such as rivers, wetlands, residual woodlands, fallow land, and hedgerows amongst fields. Harvesting rights are not codified and are vulnerable to increasing privatisation under pressures from market or population growth.

Human and natural systems are thus inseparable, and the legislative and institutional framework forms a pivot- or a bridge - between livelihoods and ecology. It also determines the direction and force of change.

Rights - allocating resources

In many dryland countries, colonialism initiated a confrontation between customary forms of natural resource tenure and alien models, which were based on European concepts of private property.[228] In the drylands of Central Asia, collectivization on a socialist model was introduced in the twentieth century. In most African drylands, the state declared its sovereignty over land, reducing customary rights to those of users, not owners. The implications were most serious for users of common land, and pastoralists in particular, who relied on local recognition of their grazing and water rights.

Box 27. Reciprocity in pastoral livelihoods in Niger

Managing natural resources through a mix of common property and private regimes, where access to pastures and water are negotiated and often depend on reciprocal arrangements, allows pastoralists to respond in a flexible and opportunistic manner to resources that are highly dispersed in time and space. Offers of reciprocity, investments in maintaining close ties with "host families" in distant lands, and careful organisation of livestock mobility allow herders to negotiate access to a wide range of resources in any given year. Therefore, besides secure resource rights over their home areas, pastoralists need flexible institutional arrangements enabling herd mobility as well as secure access to distant water and dry-season grazing. Such flexibility enables livestock to be driven to where the most nutritious and abundant pastures exist, thereby optimising weight gain and milk production in the wet season, and limiting weight loss in the dry season.

Sources: Thébaud, 2002; Turner, 1999b; Lane, 1998.

For example, the rights of pastoralists in Niger were governed by reciprocity (Box 27). But governments often treated such 'unoccupied' land as available for state appropriation or for allocation to individuals and corporations under statutory tenure. In Central Asia also, mobility was restricted, as collectives with defined territories replaced more mobile pastoral groups, with consequences for the sustainability of the pastoral system (Box 28).

The rights of farmers were more individualised and secure as they could be reasserted every year through the act of cultivation. Farmland is bounded and subject to inheritance and often exchanged. Nevertheless, across the drylands of Africa, land legislation tends to emphasise state ownership or control, with much of the population only enjoying use rights. The lack of a clear definition of what constitutes "productive use" creates opportunities for abuse, and undermines the security of land rights.[229] For example, the *Code Rural* in Niger has defined "positive" land use activities to consist largely of some form of physical or material investment (e.g. planting trees, establishing private forests, fencing off land), which is skewed in favour of agriculture and forestry.[230]

[228] (UNDP-DDC, 2001)

[229] (Cotula, 2007)

[230] République du Niger, *Code rural*, 1997.

Box 28. Institutional framework of pastoralism on the Tibetan Plateau

Collectivization was introduced to northern Tibet in 1958. Before then, tribal chiefs owned 80–90 percent of land and livestock, and poor people worked for them in a feudal relationship. Animals and land were merged to form peoples' communes, production brigades and production teams. The transition was aided by familiarity with communal herd management and risk sharing. However, major consequences for the social and cultural fabric included (a) reduced mobility enforced by smaller pasture areas, leading eventually to a threat to sustainability; and (b) a government-driven preference for sheep over *yak*, which also increased pressure on the pastures. In the early 1980s, the policy was reversed with the Household Responsibility System; livestock were turned into a private asset. After a decade, pasture land was allocated according to the numbers of household members and animals, on a 50–year lease from the Government. Government marketing monopolies were removed in the 1990s and taxation ceased to be based on the numbers of animals.

Source: Nori *et al.*, 2008.

There is no recognition of livestock mobility having 'positive' effects on the environment or on the transformation of biomass into animal products for consumption and sale.[231]

In West Africa, reinvented and adapted forms of customary tenure are applied even where they are inconsistent with legislation, because they tend to be more accessible to rural people, while accommodating the variable nature of ecosystems. As a result, several tenure systems – state, customary and combinations of both – may coexist over the same territory, resulting in overlapping rights, contradictory rules and competing authorities. This situation creates confusion and fosters tenure insecurity, which has been shown to discourage agricultural investment, undermine incentives for sustainable land management, and enable elites to grab common lands.[232] Privatization leads to individual, clearly bounded ownership, and this is preferred by farmers. But it is ill-suited to regulate the flexible, overlapping and reciprocal relations that characterize pastoral land use.

The opportunities for meeting these challenges are illustrated below, with examples from Africa.

Securing local land rights

Giving full legal recognition to local (including customary) land rights, through which most people gain access to rural land, is a key step to improved security. Land registration may (but need not) be a component of a broader strategy if customary systems have collapsed, land disputes are widespread, or in newly settled areas. Registration may also be useful in areas of high land values, such as urban and peri-urban areas and irrigated lands. A wealth of experience on how to secure local land rights is being developed in several countries, for example in the Ethiopian state of Tigray (Box 29).

Registering collective land rights may also be a cost-effective way to provide adequate tenure security, provided that group members enjoy clear rights over their plots. In Mozambique, for instance, while all land belongs to the state, "local communities" can register a collective, long-term interest and manage land rights according to customary or other local practices.[233] Several countries have made explicit efforts to protect customary land rights and provide for their registration (e.g., Uganda, Mozambique, Tanzania, Niger and Namibia).

Enabling access to appropriate systems of land dispute resolution can provide greater returns in terms of certainty and security than investing in comprehensive exercises to document everyone's land rights. Increasing the security of land transactions, particularly land rentals (fixed-rent or sharecropping contracts), is also critically important, requiring simple local documentation

Box 29. Land registration in Tigray, Ethiopia

Simple, low-cost and accessible local land records are handled by the lowest level of local government. Fees tend to be very low, the technology is very simple and the language used accessible to most rural land users. As a result, the process is transparent and accessible for most land users. However, the simple technology used does not enable documentation of the size, boundaries and location of the plots, which limits the use of the records in solving border disputes.

Source: Haile *et al.*, 2005.

[231] (Hesse and Thébaud, 2006)

[232] (Cotula *et al.*, 2006; Faye, 2008; Lo and Dione, 2000; Toulmin and Quan, 2000).

[233] (Chilundo *et al.*, 2005)

systems. New technologies such as computerized land information systems and GPS can help put in place efficient and publicly accessible land records, but they are no substitute for a locally legitimate process to adjudicate competing claims.[234]

Facilitating pastoral mobility in the Sahel

The past decade has seen a promising shift by several governments to recognize and regulate access and tenure rights over pastoral resources – first with Niger's Rural Code (1993) and then with the pastoral laws passed in Guinea (1995), Mauritania (2000), Mali (2001) and Burkina Faso (2002). Although the approaches taken by legislators vary considerably across countries, this pastoral legislation tends to recognise mobility as the key strategy for pastoral resource management – contrary to much previous legislation, which was traditionally hostile to herd mobility (Table 5).

In order to maintain or enable mobility, pastoral legislation seeks to protect grazing lands and cattle corridors from agricultural encroachment and to secure herders' access to strategic seasonal resources. The tools used range from the delimitation of pastoral lands to innovative legal concepts like the *terroir d'attache* in Niger.[235] Pastoral laws also regulate multiple and sequential use of resources by different actors (e.g., herders' access to cultivated fields after harvest), and determine the role which pastoral people can play in local conflict management.

While these laws constitute a major step forward, some problems remain. First, pastoral legislation has scarcely yet been implemented. Secondly, although some laws now recognize pastoralism as a legitimate form of productive land use (*mise en valeur*, a prior condition for protection of land rights), the pastoral application of the concept (*mise en valeur pastorale*) remains ill-defined, and generally involves investments in infrastructure (wells, fences, etc.) that are not required in the agricultural application. Finally, in most countries, other laws and institutions affect rangelands, often with contradictory or ambiguous provisions.[236]

Table 5. Key features of pastoral laws in West Africa

1. *Recognition and protection of mobility:* Pastoral Charter (Mali)	• "Throughout the country, livestock may be moved for sedentary livestock keeping, transhumant livestock keeping or nomadic livestock keeping" (Art. 14). • "Livestock mobility takes place on livestock corridors. These are local corridors and transhumant corridors" (Art. 15). • "Local government is responsible for managing livestock corridors with the help of pastoral organisations and in collaboration with all concerned stakeholders" (Art. 16). • "Any form of occupation, blockage or use of a livestock corridor or any infringement whatsoever is strictly forbidden" (Art. 17).
Pastoral Law (Mauritania)	• "Pastoral mobility is protected under all circumstances and can only be limited temporarily and for reasons of the safety of animals and crops, and this in accordance with the provisions of the law" (Art. 10).
2. *Recognition of priority use rights over resources* Rural Code (Niger)	• "Priority use rights over natural resources situated in those zones defined as "home areas" (Art. 28).
3. *Recognition of "productive" pastoral land use* Pastoral Charter (Mali)	Productive pastoral land use is defined as "the regular and long-standing use of an area for pastoral activities on public land involving customary or modern improvements and/or activities seeking to protect or restore the environment" (Art. 49).

[234] (Cotula *et al.*, 2006)

[235] Under Niger's Rural Code and its implementing regulations, the *terroir d'attache* is the area where herders spend most of the year (usually a strategic area, such as a *bas-fond* or the land around a water point), and over which they have priority use rights. Outsiders may gain access to these resources on the basis of negotiations with the right-holders.

[236] (Cotula *et al.*, 2006)

Participatory research and development involving the community and building on local knowledge, Diourbel region, Senegal. © IUCN Photo Library/*Michael Mortimore*

According to an earlier *Challenge Paper* in this series, there are six challenges for land reform in the world's drylands: (1) participation and transparency; (2) appropriate interventions in land markets; (3) a respect for customary rights; (4) recognition of multiple users' claims; (5) ensuring benefits to marginalised groups; and (6) collaboration amongst all actors.[237]

Reform - decentralising natural resource governance

Devolving responsibilities to local government bodies can strengthen local control, and provide real as well as perceived security of resource rights, provided that the bodies are truly representative and the over-arching institutional framework legitimises and upholds devolved decision making. Devolution proceeds at different speeds and to a differing extent. For example, in some countries, local governments have long enjoyed powers in natural resource management, but in Mali the policy is not yet operational. Devolution provides opportunities for legal recognition of community-based management rules that are better adapted to local environmental, social and political contexts.[238] However, it can also bring new opportunities for rent seeking and resource grabbing by local elites.[239]

Devolution of powers needs to be distinguished from de-concentration, or mere transfer of responsibilities to field units of the same administrative department. However, this can also improve local responsiveness and oversight of decision-making. Examples are the Land Commissions in Niger, the Land Boards in Uganda, and the Communal Land Boards in Namibia, where Botswana's longstanding Land Boards have been used as a model. Slow implementation of these provisions has been mainly due to a lack of human and financial capacity.

Bolivia, Ethiopia, India and Senegal have national policies of decentralization which have proceeded faster as a result of legal and institutional reforms. In Senegal the policy is 30 years old. In former centrally planned economies, progress is slower,

[237] (UNDP-DDC, 2001)

[238] (Barrow, 1996; Vogt and Vogt, 2000)

[239] (Djiré, 2007)

Box 30. Programme d'appui à la gestion de la réserve nationale de l'Aïr et du Ténéré (PAGRNAT)

PAGRNAT developed an approach to sustainable rangeland management based on customary practice that promoted livestock mobility and thereby the opportunistic tracking of resources in a highly unpredictable environment. Using the Tuareg concept of *echiwel*, the project identified up to twenty *terrain de parcours*, socially defined areas regularly used by a group of families and their livestock with priority rights of access over key resources (e.g. dry season water, grazing). The overlapping and fluid nature of these areas' boundaries as well as the practice of negotiated access by the inhabitants of the different *terrain de parcours* enabled the local population to make optimal use of the available resources and match livestock numbers to available forage in most years. The project's decision to base its operational approach on the notion of *terrain de parcours* ensured a high degree of appropriation by the local community as well as a strong basis for the design of a "model" for decentralised natural resource management and local development within the Aïr-Ténéré reserve.

Sources: PAGRNAT, 2001; 2002.

as attitudes must change and institutional inertia must be overcome.[240] Democratic reform is essential. These elements of political economy thus have a direct bearing on the empowerment of dryland resource users, and the poor in particular.

Projects such as the *Programme d'appui à la gestion de la réserve nationale de l'Aïr et du Ténéré* (PAGRNAT, Box 30), the *Projet appui à la gestion conjointe des ressources sylvopastorales* (PAGCRSP) and *Appui à la sécurisation foncière* (ASEF II) in Niger have experimented with some success in promoting livestock mobility either within pastoral areas (PAGRNAT) or between the pastoral zone and more southerly areas (PAGCRSP, ASEF II). These projects have sought to secure pastoral access and control over strategic resources (water and grazing lands, particularly in the dry season), in both the pastoral and agricultural zones of Niger. They also aimed to institutionalise decentralised management within the context of Niger's local government reform programme.

Forest law is an area where major revisions are being undertaken world-wide in the relations between users, the state and local governance institutions. The colonial inheritance in many West African countries, for example, included draconian penalties for felling trees (even on private farmland), semi-military forest police to guard reserves, and directives passed down from forest officers convinced that local land users would, unchecked, destroy the forest estate.[241] Either by lapse or reform these injunctions have largely been replaced by official recognition that farmers at least have good reason to conserve privately owned trees, claimed to have been instrumental in a remarkable recovery of indigenous agroforestry in Niger during the last decade.[242] However, pastoralists are still blamed for degrading tree stocks in common access forest reserves. There is opportunity for building forest conservation through decentralization, stakeholder negotiations, local byelaws, removal of cutting permits (except in state-owned reserves), extension services to promote agroforestry as a business, and tenure law reform.[243]

Decentralization in natural resource governance is driven by economic liberalisation as well as democratic reform, and may meet with resistance from some state interests, for example in India.[244] In the reform process, there is a risk of marginalising groups (women in particular) that are weakly represented in local power structures, and undermining multiple rights of access. There is much to be said both for and against.[245] However, the process of decentralisation offers opportunities for consolidating local democracy and downward accountability, thus progress in natural resources management may have wider benefits for society. Note that these issues are as relevant to drylands as to other ecosystems.

A thorough discussion of rights to water is beyond the scope of this paper, and the critical need for water far transcends the boundaries of dryland ecosystems. However, the rarity of surface water resources, and the dependence on well water from subsurface aquifers in dryland livelihoods ensures that water rights are integral to every regime of natural resource management. At local level, especially where dug on private land, wells could be privately owned and entered into contracts between pastoralists and farmers, as

240 (Wily 2006: p. 45)

241 (Cline-Cole, 1997)

242 (Ribot, 1995; WRI, 2008)

243 (ICRAF, 2008)

244 (Meynen and Doornbos, 2005)

245 (WRI *et al.*, 2005: pp. 70-90)

Box 31. Integrated ecosystem management in Nigeria and Niger

The project *Integrated Ecosystem Management in Shared Catchments in Niger and Nigeria* (funded by GEF/UNEP) brings together two countries (through the inter-governmental Niger-Nigeria Joint Council), four regions, six states, a larger number of departments or local governments, and several hundred villages for sharing knowledge and decision making in the area of ecosystem management. It aims to achieve sustainability at the scale of the four trans-boundary river basins linking the two countries, through a decentralised hierarchy of governing institutions. These will integrate scientific with local knowledge, ecosystem with livelihood management, and cross-border, collaborative decision making, at levels from national to local government and village areas. Provided that the temptation to find technical fixes is resisted, and genuine institutional collaboration, equitable (and effective) knowledge sharing (including due respect for local or indigenous knowledge) and adequate service provision take place, the project may succeed in pioneering more flexible pathways for development at the grassroots - of course, subject to continuity of a relatively long (8 years) funding cycle.

Source: UNEP/GEF, 2005.

among the Bambara of Mali.[246] At the opposite scale, huge state investments in irrigation, based on weak hydrological research, contributed to the degradation of wetland ecosystems in Central Asia - the Aral and Balkhash basins[247] – or fell victim to persistent drought in the Sahel – Lake Chad. Water governance can only be decentralised up to a point – that of the basin or catchment which requires data-based planning as well as local participation. At this scale, as found in Brazil, the decentralisation of natural resources management is interwoven with institutional interests up to national level, such as the municipalities and civil society groups.[248]

In the Sahel, and specifically in northern Nigeria and neighbouring Niger, major rivers flow through semi-arid ecosystems towards Lake Chad, the desert, or the ocean. Current disputes over allocating scarce water resources - among several states of Nigeria - underline the need for equity at the basin and inter-basin levels. Governance at the higher national level is required as well as decentralization at the local level. In some basins, international cooperation is also essential. Integrated ecosystem management challenges traditional structures of governance as both multi-sectoral and multi-level integrated approaches are required, contradicting the familiar patterns of sectoral administrations and hierarchical authority (Box 31).[249]

Governance of natural resources is a critical determinant of outcomes wherever natural ecosystems are co-evolving with human or social systems.[250] Human agency can accelerate or postpone natural cycles of ecosystem breakdown and recovery, and it has been argued that the idea of 'ecosystem health' should be extended to a broader concept of 'eco-cultural health' which includes 'a design for re-integrating society with nature through a shift in values, and through more adaptable and holistic governance systems'.[251] Dryland ecosystems are subject to unpredictable perturbations (variability) and their successful management depends on matching the resilience of the ecosystem with a corresponding adaptive capacity in the human (social and economic) system. To an extent, such resilience was achieved (at considerable social cost) in pre-modern livelihood systems such as those of the Sahel. The challenge faced today consists in building structures that can accommodate increasing competition for ecosystem services (cultivable land, rangeland, trees, water, wild products), and new opportunities (such as biofuels, carbon capture). Multiple claims must be dealt with, and more flexible concepts of rights to territory must be promoted. Before they lose out to competing and private sector interests, poor people (and the less poor) in the drylands need legal empowerment (or, in the light of history, re-empowerment).[252]

In an agro-pastoral Sahelian context, some of these dilemmas have been resolved, and competing interests reconciled, by means of negotiated local conventions. Local conventions are community-based agreements on the management of shared natural resources, which are negotiated by the users, often but not necessarily with support from external agencies (Box 32). Sometimes formalised through local government byelaws, they regulate (for example) pastoral mobility and access to water points, dry-season pastures and post-harvest fields (where the most valuable fodder is

[246] (Toulmin, 1992)

[247] (CAREC, 2003)

[248] (Brannstrom, 2005)

[249] (UNEP/GEF, 2005)

[250] (Gunderson *et al.*, 1995)

[251] (Rapport, 2009)

[252] (CAREC, 2005; McAuslan, 2006)

Box 32. Local convention at Takieta, Niger

The aim of the *convention* is to achieve equitable, rights-based, collaborative and sustainable common property resource management in a formerly degraded forest reserve of 6,720 ha, with 23 villages and a population of 38,000 sharing access. The project was carried out in the context of Niger's Rural Code (from the 1990s) and its decentralisation process (from 2004). With support from an NGO and the Government, local users negotiated a set of rules and institutions that enable sustainable resource use, and peaceful coexistence among competing resource users. These users include farmers, pastoralists, wood-cutters, and harvesters of wild products. Takieta has been managed successfully for a decade, and the experience has led to four other local forest areas being brought into parallel processes of negotiation, positively affecting the lives of an estimated 80,000 people.

Source: Vogt and Vogt, 2000.

found). In such negotiations, mobile pastoralists, being unable to assert their rights by cultivation, may need support through their own institutions to participate on an equal footing.[253] A prior condition of all such consensus-building negotiations is that government should relax its control over natural resources and place its trust in local users' institutions.

The need for local consensus on the governance of natural resources is demonstrated in several conflict situations in African drylands, and most conspicuously in Darfur. An institutional failure has exacerbated the effects of chronic under-investment, ethnic and political factors, variable rainfall, and a transition in livelihoods involving migration and urbanization.[254] Experience in the Kenya drylands shows that conflict reduces the effectiveness of existing adaptive strategies. Adaptation needs to be taken up as a development objective and funded across ministry boundaries.[255] At an international level, the achievement of the MDGs in the drylands is a challenge for global governance.[256] Governance of the natural resource base of economic and social development is accorded a critical role by the Economic Commission for Africa.[257]

Risk – managing uncertainty

In the context of global exposure to disasters, 'appropriate governance is fundamental if risk considerations are to be factored into development planning and if existing risks are to be successfully mitigated.'[258] This has particular relevance to drylands. In a variable climate, institutions are needed to protect, or to limit the loss of, critical productive assets such as livestock, ploughs or land during bad years, while enabling communities to rebuild their capital and productive livelihoods once the crises have passed. Such institutions minimise chronic destitution and the costs of lost production. They must also support adjustment to economic, political or social shocks, protect assets and enable diversification and mobility. Providing institutional incentives to manage expected but unpredictable cycles of "boom and bust" requires considerable innovation and capacity. This principle is illustrated below by two examples: first, insurance for herders, and second, for farmers.

Insurance for herders

WoDaaBe herders in Niger, as well as using mobility to insure themselves against rainfall and pasture variability, also seek to maximise their herds, and their returns, in good years. The size of herd thus represents the risk profile of a pastoral family. Families with larger herds spread risk by splitting them into smaller management units, and can make loans to kin or friends, thereby building social capital. In a larger herd there is likely to be greater diversity in age and sex distribution, which determines how quickly the herd will recover after losing animals in a drought (Box 33). Such institutions - embedded in the community and its knowledge of natural resources - perform an essential function in drylands and should never be put under threat by development innovations. Examples, presently decreasing owing to poverty, are habanae (animals loaned between friends to help re-stocking) and their equivalents among other peoples - *ewolotu* (Maasai) and *debere* and *busaa gonofa* (Boran). Perhaps the breeding of cattle in matriarchal lines, to ensure high performance in a variable environment rather than mere productivity, is in a sense a WoDaaBe 'institution'.[259]

[253] (WISP, 2007a)

[254] (Bromwich *et al.*, 2007)

[255] (Eriksen *et al.*, 2006)

[256] (Bird, 2008; Dobie and Goumandakoye, 2005)

[257] (ECA, 2008)

[258] (UNDP, 2004)

[259] (Krätli, 2008)

Box 33. WoDaaBe insurance strategies in Niger

A balanced herd structure is a critical insurance strategy among these mobile herders. Adult cows are needed to produce milk, in the short term, and calves that will ensure the future survival of the herd and thus of the family. Adult steers are needed for sale to buy grain and other foods and services, or for major ceremonial purposes central to maintaining social capital. A bull is needed to inseminate the cows. Heifers are needed to replace the cows. Young steers are fattened for future sale. Having access to adult males after a drought, when livestock prices are high, allows the WoDaaBe to preserve breeding females for milking and calving. By contrast, a sedentary Fulani herd was found to have fewer animals and an imbalanced herd structure, dominated by females, at the end of a drought. The herders were therefore forced to sell female stock, thereby compromising ability to reconstitute the herd.

Source: Thébaud, 2002.

Insurance for farmers

Farmers also maintain webs of social claims and obligations at the community level. The difference with herders is that the social fabric is proving weaker in the face of new opportunities to pursue wealth through markets by individual strategies. Moreover, dependence on annual crops rather than livestock increases vulnerability to food insecurity in the short term for claimants and benefactors alike. During the Sahel drought of the early 1970s, farmers in severely affected areas in northern Nigeria bewailed the inability or unwillingness of local patrons to offer assistance to their poor clients; instead they appealed to the government. Wisely unwilling to wait, however, they set about an elaborate framework of strategies for finding alternative incomes.[260]

While the utility of seasonal weather forecasting and early warning has provided a natural focus for experiments in insurance provision, crop failures may extend over large geographical areas and call for backstopping by government, regional or international organisations. The Ethiopian government, in partnership with the World Food Programme, has introduced a scheme whereby the government pays an annual premium to a private sector company which undertakes to pay compensation to farmers in the event of drought, as measured by a weather index.[261] A scheme for groundnut farmers is being tried in Malawi (Box 34). Groundnut is a market crop, and food crop production will not benefit directly. However, as markets become more pervasive in drylands (see Chapter 6), the benefits of insuring market production will impact more widely on livelihoods.

Financial insurance schemes provide an opportunity to involve the private sector in dryland development. In several schemes that now operate, a Weather Index is used to trigger payouts, or a related index such as animal mortality. In Mongolia, two insurance products can now protect herders.[262] The base insurance product is a commercial policy sold and serviced by insurance companies. The product pays out when the mortality rates in the region exceed a specified trigger. The maximum payment is at an agreed level. If losses in the region exceed this level, the government's Disaster Response Product (DRP) compensates all herders (including those who don't buy the private insurance).

Micro-credit can be used to flatten out seasonal fluctuations in the prices of food and livestock, and especially for reducing the impact of drought on women, who (in several dryland countries) bear much of the responsibility for feeding children.

Box 34. Groundnut insurance in Malawi

Opportunity International Bank of Malawi and Malawi Rural Finance Corporation give groundnut farmers loans for high-yielding certified seed, as long as they buy weather insurance (provided by the Insurance Association of Malawi). If there is a drought that triggers a payout, that money will be paid directly to the bank in order to pay off the farmer's loan. If there is no drought the farmer will benefit from selling the higher value production. This arrangement has allowed farmers to access new finance. Reducing risk exposure can give producers confidence to invest in inputs and strategies for higher returns in other years.

Source: Weather Index Insurance Malawi www.microinsurancecentre.org

[260] (Mortimore, 1989; Watts, 1983)

[261] (Barrett *et al.*, 2007)

[262] (Mahul and Skees, 2006).

Diversification as insurance

To treat livelihood diversification solely as a risk-spreading strategy would be to ignore the multiple benefits it offers to both individuals and families as a way of escaping poverty, for women as well as for men. Diversity characterises not only the sources of alternative incomes and the market systems in which they are sought, but also the age, gender, skills, education, and motivation of participants; the places, seasons and times of participation; the entry costs or barriers, remuneration or profits, and sharing of benefits. However, two generalisations may be made.

First is the commonly observed intensification of the search for alternative strategies when drought destroys incomes from crops or livestock. Examples from West Africa include investment by farmers in livestock (taking advantage of depressed prices), and increased agro-pastoralism; accelerated off-farm employment and short- or long-term migration; and investments in small-scale irrigation. Second is a tendency for governments to ignore this 'informal' but market-based sector in policy making, or to obstruct it through controls on movement by rural people to cities.

Dryland communities are highly differentiated. Some are heavily dependent on natural resources, others less so. Some make good use of a diverse portfolio of income-generating activities; others are less successful, even falling into destitution. Environmental variability, coupled with rising population levels and increasing competition for access to natural resources, adds urgency to the need for institutions that can promote viable alternative incomes and investment options, and not only for the better off. As argued in Chapter 6, new market opportunities both for natural resource products and for other activities offer alternative development pathways for dryland communities. Targeted interventions, such as micro-credit provision for women's income-generating activities (e.g., CARE-Niger), are practicable and appropriate.

In this chapter we have argued that the three dominant livelihood systems in drylands - farming, pastoralism and wild harvesting - are not inevitably bound to a course of ecological destruction, as commonly suggested, but contain within themselves the seeds of sustainability and well-being. The governance and institutions of dryland systems are critical. And contrary to what was often supposed in the past, rather than coercion, prevention and direction from the top, dryland communities need to have their autonomy, equity and capacities restored through democratic governance and participatory and collaborative institutions, based on efficient sharing of knowledge (and recognition of rights).

Resilience – building sustainable livelihoods

Scope has been found by development agencies for new institutional forms - often built on pre-existing local institutions - for enhancing poor peoples' livelihoods through sustainable use of ecosystem services. The World Resources Institute reports successes in a number of dryland countries.[263] Many are covered by the term "Community based natural resource management" (CBNRM). This approach is useful for restoring and managing common pool resources (e.g., forests in Senegal; nature conservancies in Namibia). Another approach is that of "integrated watershed planning", which is based on hydrological principles implemented through community organisations (e.g., as widely applied in India, also see Box 31). Women's groups, often building on local rotating credit groups, have taken on new levels of activity in many African countries.

The old model of an intervention led by new technology to address supply constraints has thus given way to a broader conception of an alliance between civil society, CBNRM groups (including women's groups), local government and the private sector to catalyse action in natural resource management, where inertia had previously been experienced in mobilising the existing institutions.[264] Community groups can be seen not only as executing bodies, but as channels for dialogue on development initiatives, for co-learning, idea-sharing and partnerships with external agencies, as has been demonstrated in Zimbabwe.[265]

A key to success in institutional development in natural resource management is to create or improve household incomes, whether from crops and livestock, or from wild harvesting of products such as wood fuel or charcoal, timber, fibre, food and fodder, and medicines. For example, income enhancement has resulted from forest restoration by communities in Tanzania (Box 35). But income is not the whole story. Collaborative natural

[263] (WRI, 2008)

[264] (UNCCD, 2006)

[265] (Nyoni, 2008)

Box 35. Forest restoration and improved incomes in Shinyanga, Tanzania

The practice of *ngitili* grazing and fodder reserves - which could be private or communal - declined in the later twentieth century owing to cash cropping , declining fertility, increasing numbers of livestock, demand for wood fuel, and the villagization policy of the Tanzanian Government. A revision of forest policy in 1998 restored participatory management and decentralization, and a government conservation and restoration programme was implemented among the Sukuma people, under whose collaborative management *ngitili* are now providing livelihood benefits including sustainable fodder and fuel, access to non-timber forest products, reduced drought risk, and amenity value. Biodiversity is protected.

Sources: Barrow and Mlenge, 2001; WRI, 2005.

resource management can correct imbalances caused by the substitution of weakly regulated 'capitalism' for socialist collectivization policies. In Central Asia, with an ecology very different from that of Tanzania, co-management is restoring both incomes and ecosystem services in Mongolia (Box 36) and northern Tibet.[266]

Civil society institutions can facilitate initiatives addressing the needs of minority groups and promoting greater equity. For example, a Pastoral Women's Council in Tanzania promotes the interests of women and children among the Maasai, by facilitating their access to education, health and economic empowerment, including access to livestock.[267] For conservation purposes, experiments in Niger and northern Nigeria suggest that external facilitation and knowledge inputs are enough to provoke community action to save valuable biodiversity.[268] Where small businesses are under-developed, as in Caspian coastal communities of the former Soviet Union, credit institutions, capacity building and grants may stimulate growth in a market economy.[269] On the other hand, where commercial activity is pervasive, such as in the Sahel, new institutional initiatives may still contribute to market chain development.[270]

The diversity of these examples indicates that institutional development must be crafted to fit into local circumstances and there is no universal prescription. In particular, indigenous institutions can still provide a sound platform for development.

Conclusion

There are many opportunities for adaptive institutional development in drylands to strengthen rights, transact reform, manage risk and increase resilience in variable environments. New partnerships between communities, governments and NGOs making the best use of both local and external knowledge are already being tried in a number of countries. There is a role for a regulated private sector. Experience gained from pilot and innovative schemes must be shared, tested and modified to 'scale up' to wider applications. Institutions work within a policy framework, and this framework must be negotiated democratically. Hence the importance of such concepts as empowerment, participation, and ownership. However, much more is needed than a mere restoration of the *status quo*. Contemporary trends, stresses and opportunities need to be better understood, and the knowledge shared more effectively with stakeholders in policy processes, from the grass roots to national legislatures.

Box 36. Revival of pastoral community co-management in Mongolia

Huge losses of livestock during the dzud (heavy snow) and drought of 2000–1 provoked a popular renewal of pastoral communities based on traditional practices to regulate access to pasture and water, which had broken down during the 1990s after the collapse of socialist collectives (75 percent of Mongolia is officially classified as degraded). Donor support assisted the dissemination of value-adding activities such as camel wool spinning, dairy processing and felt making. Between 2002 and 2003, the income from livestock products in 72 sampled community households doubled, whereas that of non-community households changed little. More than half of community households reported environmental improvement resulting from community action. Improved pasture management, increased mobility, and conservation of bushes and trees were reported.

Source: Undargaa, 2006.

266 (ICIMOD and WISP, 2007)

267 (Ngoitiko, 2008)

268 (Mortimore *et al.*, 2008)

269 (CAREC, 2003; CAREC, 2006)

270 (Bolwig *et al.*, 2009)

Naibunga Conservacy, northern Kenya. © *Jonathan Davies*

CHAPTER 8
Action for the world's drylands

The Millennium Development Goals cannot be met, nor sustainable ecosystem management achieved, unless drylands are brought back into the mainstream of global development. This Challenge Paper shows the opportunities that exist for achieving these aims, thereby benefiting 41.3 percent of the earth's land surface and 35.5 percent of its population.[271]

Commonly held presumptions about poverty-environment links require review, and a new set of global drivers of change needs to be identified. A new dryland paradigm should be built on the resources and capacities of dryland peoples, on new and emergent economic opportunities, on inward investment, and on the best support that dryland science can offer.[272]

Action is necessary

Such a vision for drylands is now a global (not a local) responsibility. A new interlocking of climatic and geo-political factors means that drylands cannot be treated any longer as poor, remote, largely self-subsistent areas and left to their own devices.

- Poor people are expected to bear disproportionately the costs of climate change. The drylands are home to a very large share of the world's poor. Within the framework of the MDGs, poor people's rights to development are an international obligation.
- The costs of adaptation to climate change now are far exceeded by the costs of repairing the damage later,[273] and as the drylands are already affected by climatic variability (both floods and droughts), a sound adaptation strategy is essential.
- Land cover (woodland, cultivation, deserts) and surface conditions (temperature, moisture) in drylands are among the drivers of global circulation which determine the climates of rich 'Northern' countries as well as poor 'Southern' ones.
- Because of their extent, dry woodlands, grasslands and farmlands can capture significant quantities of CO_2. On the other hand, large scale forest burning may contribute to global carbon emissions. The management of dryland ecosystems is therefore integral to global sustainability.
- Dryland ecosystems are home to faunal and floral biodiversity which supports a wide range of ecosystem services (food, fodder, medicines, nutrients, materials), provides a safety net in times of food scarcity, conserves seed banks to support agro-diversity, and attracts tourism.
- Unfair global markets (exclusion, subsidies and dumping) can undermine the profitability of dryland production systems, depriving them of investment incentives.
- Economic inequality is threatening to destabilise international relations, not least in growing flows of migrants from dryland countries trying to gain access to the wealth of rich nations.
- Long-term internal conflict, focused on the control of natural resources, is tearing some dryland countries apart, and has implications for international security.
- Global food security is more delicately poised than it has been since World War II, with rising food commodity, fuel and input prices. The cost of food aid is increasing, and far exceeds the cost of measures to improve local production. However food production is already competing with biofuels for land and capital in some dryland countries.

Building blocks for a dryland strategy

A strategy is needed that will achieve three aims: enhancing the economic and social well-being of dryland communities, enabling them to sustain their ecosystem services, and strengthening their adaptive capacity to manage environmental (including climate) change. This is a developmental rather than a solely technological pathway, based on the principle that sustainability is conditional on (appropriate) development. The following five building blocks are proposed:

1. Upgrading the knowledge base

Drylands suffer from an exceptionally wide gulf between knowledge and policy or practice, as shown in many interventions that have not succeeded.

---

271 Drylands include the following agro-climatic zones: dry sub-humid, semi-arid, arid and hyper-arid (deserts). We exclude the USA, Canada, Australia, smaller industrialised areas and arctic regions.

272 (Menon, 2008)

273 (Stern, 2007)

The simplistic assumptions used to support some dryland development interventions and conservation efforts in the past need to give way to more accurate, complex, risk-aware and participatory models and strategies, based on the full recognition of indigenous rights. Dryland ecosystems stand in need of improved understanding (of, for example, seasonality, variability, ecosystem services such as water, and human systems).

- Local knowledge – often ignored in the past – needs better recognition, sensitive use, and scientific strengthening. It is at community level that decisions on resource use and regulation are made and sustained.
- Research-based knowledge, including climate change, adaptation, and sustainable land management, needs to be effectively disseminated both at policy and community levels.
- Knowledge partnerships bringing together communities, policy makers, institutional and commercial stakeholders, and scientists need to be constructed to reflect the political-environmental relations of places and issues.
- Community-based learning processes, and natural resource management contracts (such as 'local conventions'), need setting up and further development in more drylands.

Emphasis is needed on knowledge use, which has been relatively neglected compared with knowledge generation. However, using new knowledge is not straightforward, and investment is needed to promote new awareness of dryland ecosystem functioning into the places where it is needed, to systematize development experience, and to share the ownership of knowledge among a wider range of stakeholders.

2. Re-evaluating and sustaining dryland ecosystem services

The true economic value of some ecosystem services is not adequately recognised in national accounts and this contributes to the neglect of drylands in economic planning and service provision.

- The supporting services provided to agricultural production by the ecosystem (soil fertility and soil moisture in particular, which includes input from scarce and valuable wetlands) are normally ignored while input costs are always factored in to agricultural economic analysis. This means that nature's contribution to these basic activities, and the costs of husbanding them sustainably, are under-estimated. Furthermore, the true value of agriculture is in subsistence production as well as in sales.
- The valorisation of dry ecosystems by pastoral herding is under-estimated because only sales of livestock products are normally on record, while natural pastures support livelihoods based on breeding as well as marketing aims.
- Forests have value beyond that of sold timber, fuelwood or charcoal. But even these values are poorly documented (and sometimes illegally gained), though in dryland countries, fuelwood and charcoal may provide up to 80 percent of energy needs. Trees are grown both on farms and in natural woodland. Non-timber forest products (NTFPs) include a huge range of useful products that are harvested from trees and shrubs both in natural woodland and on farms. In addition to their value at home, wild products are finding new and sometimes little known markets. Recent studies in Africa have begun to provide estimates of the value of these provisioning services.
- Solar energy in inhabited drylands is second only to that in the hyper-arid (desert) biome.
- Tourism in drylands is based on the cultural services (scenery, animals, etc.) provided by the ecosystem. In those countries having a tourism sector, most revenues are earned in the drylands.
- The regulating services of dryland ecosystems include such functions as water filtration and sub-surface storage, all the more valuable in a seasonal regime with no rainfall for half or more of the year.
- The biodiversity of dryland ecosystems is greater than commonly supposed, and valued by local communities. The resilience of dryland ecosystems in variable and uncertain conditions has direct value for local communities, for example in providing famine foods.

These values play critical roles in first, the rationales that underlie land use systems such as mobile livestock herding and extensive farming, second in local knowledge and innovation, and third in adaptive capacity to changing environments. These should be better understood as resources on which to build, rather than impediments to remove, in the furtherance of sustainable development. Since many ecosystem services are obtained from places legally or presumed to be of open access, appreciating market values may provoke destructive exploitation. So an additional but compelling reason for a correct valuation is to provide a sound basis for conservation policies and regulation.

3. Promoting public and private investment in drylands

It is incongruent that public policy should ignore investment potential in drylands while poor resource users struggle to invest their own small resources in sustainable management of privately owned land and other natural resources. But this is often so. Yet there is positive evidence of benefits from investment, such as:

- Positive impacts on poverty of public investments in infrastructure and services in India and China.
- Satisfactory economic rates of return at the project level (several projects in African drylands).
- Successful and growing marketing of wild resources, locally, nationally and overseas (e.g., southern Africa).
- Tourist industries representing both public and private investment in dryland countries.
- Poor people's investments of their labour and skills (where finance is scarce), to maximise the productivity of farms, herds or farm trees. Large-scale, commercial, private sector investments are problematic in the uncertain environments of drylands; but small-scale producers can reap benefits from intermittent, incremental micro-investments.

The question of investment needs to be scrutinised from a range of standpoints – those of poor resource users, private commercial enterprise, national productivity, governance and sustainability - and not merely through an accounting framework. Most forms of micro-investment (other than those financed by development programmes) have been officially ignored. Policy incentives for investment should be approached from the standpoint that the greater the value of an asset, the more likely it is that right-holders will wish to sustain it.

4. improving access to profitable markets

Markets and value chains are undergoing a transformation in many drylands. In place of a 'colonial' export market added on to a much larger subsistence or non-market sector, new market relations are penetrating every corner of the human-ecological system.

- Export markets for the old dryland products – cotton, groundnuts, hides and skins – although still important to some countries, are stagnating or declining under the impact of low world prices, tariff barriers, substitutes, and increasing costs of production. For such countries, diversification is a top priority.
- Rapidly expanding internal or regional markets for food commodities, driven by urbanization and rising incomes, are increasing demand for staple grains, meat and dairy products, and for imported wheat or rice. Urban provisioning drives both local and cross-border trade. Levels of participation are increasing, and these markets are becoming more efficient.
- New or rapidly expanding niche markets for natural products, as well as some hitherto neglected indigenous crops, are providing additional impetus to urban and export trade.
- Entirely new formal value chains are emerging, linking small-scale collectors or growers with bulking, processing and packaging companies. Some products (such as flowers, green beans) can bear the capital and input costs of large-scale greenhouses, irrigation, and air freight to Europe, increasing the scale of financial transactions in drylands by an order of magnitude.
- A new market for biofuels is imminent in drylands, though with major drawbacks: first, profitability seems most likely to be assured by large-scale methods rather than small-scale outgrowers, and second, likely competition with food crops for scarce land. Equity and food security must be assured.
- Markets for labour, land, water and other natural resources, finance, inputs, knowledge and services are evolving in response to the monetization of economies and a policy environment of trade liberalisation.

Transport and communications infrastructure, locally responsive regulating institutions, adequate information systems, and fiscal stability are all needed to promote access to growing markets. Some drylands are remote from cities or ports, but this obstacle has been overcome before. Value chains interact with ecosystems in ways specific to each commodity. To discourage destructive exploitation, appropriate institutions are required.

5. Rights, reform, risk and resilience

Institutional frameworks of ecosystem management in the drylands face new challenges, thrown up by current demographic, economic, political and social trends. Progress has been made, however:

- Governments have under-estimated the difficulties of direct interventions (such as new land legislation and nationalisation) in rights to land and other natural resources. These have not always worked efficiently or equitably, and some have met with resistance. Meanwhile, customary rights in some countries have evolved spontaneously. More flexible models are available that better accommodate customary practice with livelihood security.
- Decentralisation of natural resources governance has progressed in many countries, especially in West Africa. Increased local voice is consistent with the movement towards democracy. Within this framework, prototypes have been developed to ensure equity and sustainability in the use of ecosystem services, and these can be further extended.
- Indigenous methods of managing risk, including the rational practices of herd mobility in pastoral systems and of accessing alternative incomes in urban or humid areas, must be respected and protected. Both local institutions and national policies can underwrite vulnerable livelihoods, for example, through insurance for herders and farmers, promotion of economic diversification, support for community-based management and for targeted assistance.
- Resilience is needed, not only internally, but in managing relations between drylands and the rest of the world. Variability in international relations and markets, for example, adds to that of the environment. In place of an inequitable dependency, improved local autonomy is the best foundation for a viable development pathway in the longer term.

References

Adams, W.M. and Jeanrenaud, S.J. 2008. *Transition to sustainability: towards a humane and diverse world.* IUCN, Gland, Switzerland.

Adeel, Z., Safriel, U., Niemeijer, D. and White, R. 2005. *Ecosystems and human well-being: Desertification synthesis. Millennium Ecosystem Assessment.* World Resources Institute, Washington, DC.

Adejuwon, J.O. 2005. Food crop production in Nigeria: I. Present effects of climate variability. *Climate Research* **30**(1): 53-60.

Adejuwon, J.O. 2006. Food crop production in Nigeria. II. Potential effects of climate change. *Climate Research* **32**: 229-245.

African Union. 2006. *Abuja declaration on fertilizer for the African green revolution.* African Union Special Summit of the Heads of State and Government, Abuja, Nigeria (13 June, 2006).

Anderson, S., Morton, J. and Toulmin, C. (forthcoming 2009). Climate change for agrarian societies in drylands: implications and future pathways. In: Mearns, R. and Norton, A. (eds.) *Social dimensions of climate change: equity and vulnerability in a warming world.* The World Bank, Washington, DC.

Ariyo, J.A., Voh, J.P. and Ahmed, B. 2001. *Long-term change in food provisioning and marketing in the Kano Region.* Drylands Research Working Paper 34. Drylands Research, Crewkerne, UK.

Aubréville, A. 1949. *Climats, forêts et désertification de l'Afrique tropicale.* Société d'Editions Géographiques, Maritimes et Coloniales, Paris, France.

Ba, C.O., Bishop, J., Deme, M., Diadhiou, H.D., Dieng, A.B., Diop, O., Garzon, P.A., Kebe, M., Ly, O.K., Ndiaye, V., Ndione, C.M., Sene, A., Thiam, D. and Wade, I.A. 2006. *The economic value of wild resources in Senegal. A preliminary evaluation of non-timber forest products, game and freshwater fisheries.* IUCN, Gland, Switzerland.

Bai, Z. G., Dent, D.L., Olsson, L. and Schaepman, M.E. 2008. *Global assessment of land degradation and improvement: 1. Identification by remote sensing.* World Soil Information Report 2008/01. FAO/ISRIC, Rome, Italy/Wageningen, The Netherlands.

Barbier, E. B., Baumgärtner, S., Chopra, K., Costello, C., Duraiappah, A., Hassan, R., Kinzig, A., Lehman, M., Pascula, U., Polasky, S. and Perrings, C. 2009. The valuation of ecosystem services. In: Naeem, S. (ed.) *Biodiversity, ecosystem functioning, and human wellbeing. An ecological and economic perspective.* Oxford University Press, New York, USA. pp. 248-262.

Barrett, C.B., Barnett, B.J., Carter, M.R., Chantarat, S., Hansen, J.W., Mude, A.G., Osgood, D., Skees, J. R., Turvey, C.G. and Ward, M.N. 2007. *Poverty traps and climate risk: Limitations and opportunities of index-based risk financing.* IRI Technical Report No. 07-02. International Research Institute for Climate and Society, Colombia University, New York, USA.

Barnett, A. and Whiteside, A. 2002. *AIDS in the twenty-first century: disease and globalisation.* Palgrave Macmillan, Basingstoke, UK.

Barrow, E.G.C. 1996. *The drylands of Africa. Local participation in tree management Nairobi.* Initiatives Publishers, Nairobi, Kenya.

Barrow, E. and Mlenge, W.C. 2001. Case study 2. Forest restoration in Shinyanga, Tanzania. In: Fisher, R.J., Maginnis, S., Jackson, W.J., Barrow, E. and Jeanrenaud, S. *Poverty and conservation: Landscapes, people and power.* IUCN Forest Conservation Programme, Landscapes and Livelihoods Series No 2. Gland, Switzerland. pp. 61-67.

Behnke, R.H., Scoones, I. and Kerven, C. (eds.) 1993. *Range ecology at disequilibrium: New models of natural variability and pastoral adaptation in African savannas.* Overseas Development Institute, London, UK.

Ben Mohammed, A., Van Duivenbooden, N. and Abdoussallam, S. 2002. Impact of climate change on agricultural production in the Sahel. Part I: Methodological approach and case study for millet in Niger. *Climatic Change* **54**: 327-348.

Bennett, B. 2006. *Natural products: the new engine for African trade growth. Consultancy to further develop the trade component of IUCN's Natural Resources Enterprise Programme.* Natural Resources Institute, Regional Trade Facilitation Programme. Pretoria, South Africa.

Berkes, F., Colding, J. and Folke, C. 2000. Rediscovery of traditional ecological knowledge as adaptive management. *Ecological Applications* **10**(5): 1251-1262.

Bester, J., Matjuda, L. E., Rust, J. M. and Fourie, H. J. 2001. Nguni: a case study. In: *Proceedings: community-based management of animal genetic resources.* Food and Agriculture Organisation of the United Nations, Rome, Italy.

Bird, A. 2008. *MDGs and the environment: are environmental institutions fit for purpose?* Overseas Development Institute, London, UK.

Boko, M., Niang, I., Nyong, A., Vogel, C., Githeko, A., Medany, A., Osman-Elasha, B., Tabo, R. and Yanda, P. 2007. Africa. In: Parry, M.L., Canziani, O.F., Palutikof, J.P., van der Linden, P.J. and Hanson, C.E. (eds.) *Climate change 2007. Impacts, adaptation and vulnerability. Contribution of Working Group II to the Fourth Assessment Report of the Intergovernmental Panel on Climate Change.* Cambridge University Press, Cambridge, UK. pp. 433–467. www.ipcc.ch/

Bolivia. 2006. *Mechanismo de adaptación al cambio climatico.* Programa Nacional de Cambios Climatico, Ministry of Development Planning/PNCC, La Paz, Bolivia.

Bolwig, S., Cold-Ravnkilde, M., Rasmussen, K., Breinholt, T. and Mortimore, M. 2009. *Achieving sustainable natural resource management in the Sahel after the era of desertification. Markets, property rights, decentralisation and climate change.* Discussion Paper submitted to TAS/Ministry of Foreign Affairs of Denmark. DIIS Working Papers. Danish Institute for International Studies, Copenhagen, Denmark.

Bonkoungou, E.G. 2001. *Biodiversity in the drylands: Challenges and opportunities for conservation and sustainable use. Challenge Paper.* The Global Drylands Initiative, UNDP Drylands Development Centre, Nairobi, Kenya.

Bossio, D. and Geheb, K. (eds.) 2008. *Conserving land, protecting water.* CABI International, Wallingford, UK.

Brannstrom, C. 2005. Decentralising water resource management in Brazil. In: Ribot, J.C. and A. Larson, A. (eds.) *Decentralisation through a natural resource lens.* Routledge, London, UK. pp. 214-234.

Breman, H. and de Wit, C.T. 1983. Rangeland productivity and exploitation in the Sahel. *Science* **221**: 1341-1387.

Brock, K. and Ngolo, C. 1999. *Sustainable rural livelihoods in Mali.* IDS Research Report 35. Institute of Development Studies, Brighton, UK.

Bromwich, B., Adam, A.A., Fadul, A.A., Chege, F., Sweet, J., Tanner, V. and Wright, G. 2007. *Darfur: relief in a vulnerable environment.* Tearfund, Teddington, UK.

Brouwer, J. 2008. *The importance of within-field soil and crop growth variability to improving food production in a changing Sahel.* IUCN Commission on Ecosystem Management, Gland, Switzerland.

Brown, O., Hammill, A. and McLeman, R. 2007. Climate change as the 'new' security threat: implications for Africa. *International Affairs* **83**(6): 1141-1154.

Bryceson, D.F. 2002. The scramble for Africa: Reorienting livelihoods. *World Development* **30**(5): 725-739.

Burton, I. 2001. *Vulnerability and adaptation to climate change in the drylands. Challenge Paper.* The Global Drylands Initiative, UNDP Drylands Development Centre, Nairobi, Kenya.

CAREC. 2003. *Programme specific actions on improvement of ecological and social situation in Aral Sea Basin (2003-2010).* Environmental Block, CAREC (Regional Environmental Centre for Central Asia), Almaty, Kazakhstan.

CAREC. 2005. *Examples of implementation. The 'Convention on Access to Information, Public Participation in Decision-making and Access to Justice in Environmental Matters' in Central Asia.* CAREC (Regional Environmental Centre for Central Asia), Almaty, Kazakhstan.

CAREC. 2006. *Progress review on education for sustainable development in Central Asia.* CAREC (Regional Environmental Centre for Central Asia), Almaty, Kazakhstan.

Chilundo, A., Cau, B., Mubai, M., Malauene, D. and Muchanga, V. 2005. *Land registration in Nampula and Zambezia Provinces, Mozambique.* International Institute for Environment and Development, London, UK.

Chuluun, T. 2008. Adaptation strategies of pastoral communities to climate change in central mountainous region of Mongolia. *Newsletter of the International Human Dimensions Programme on Global Climate Change* **2**: 53-58.

CIA. 2008. *World factbook: Senegal.* Central Intelligence Agency, Washington, DC.

Cline-Cole, R.A. 1997. Promoting (anti)social forestry in northern Nigeria? *Review of African Political Economy* **74**: 515-536.

Cline-Cole, R. 1998. Knowledge claims and landscape: alternative views of the fuelwood-degradation nexus in northern Nigeria? *Environment and Planning D: Society and Space* **16**(3): 311-346.

Cline-Cole, R.A., Falola, J.A., Main, H.A.C., Mortimore, M., Nichol, J.E. and O'Reilly, F.D. 1990. *Wood fuel in Kano.* United Nations University Press, Tokyo, Japan.

Cooper, P.J.M., Dimes, J., Rao, K.P.C., Shapiro, B., Shiferaw, B. and Twomlow, S. 2008. Coping better with current climatic variability in the rain-fed farming systems of sub-Saharan Africa: an essential first step in adapting to future climate change? *Agriculture, Ecosystems and Environment* **126**(1-2): 24-35.

Cotula, L. 2007. *Legal empowerment for local resource control: securing local resource rights within foreign investment projects in Africa.* International Institute for Environment and Development, London, UK.

Cotula, L., Toulmin, C. and Quan, J. 2006. *Better land access for the rural poor: Lessons from experience and challenges ahead.* International Institute for Environment and Development, London, UK.

Cotula, L., Vermeulen, S., Leonard, R. and Keeley, J. 2009. *Land grab or development opportunity? Agricultural investment and international land deals in Africa.* IIED/ FAO/ IFAD, London, UK.

Cour, J.M. and Snrech, S. (eds.) 1998. *Preparing for the future: A vision of West Africa in the year 2020: Summary report of the West Africa long-term perspective study.* Club du Sahel, OECD/OCDE, Paris, France.

de Oliveira, T., Duraiappah, A.K. and Shepherd, G. 2003. *Increasing capabilities through an ecosystems approach for the drylands. Challenge Paper.* The Global Drylands Initiative, UNEP Drylands Development Centre, Nairobi, Kenya and UNEP, Nairobi, Kenya.

DFID. 2006. *Chief Scientist's climate change and Africa report.* Unpublished report, Department for International Development, London, UK.

Djiré, M. 2007. *Land registration in Mali - no land ownership for farmers?* IIED Issue Paper 144. International Institute for Environment and Development, London, UK.

Djurfeldt, G., Holmén, H., Jirström, M. and Larsson, R. 2005. *The African food crisis. Lessons from the Asian Green Revolution.* CABI Publishing, Wallingford, UK.

Dobie, P. 2001. *Poverty and the drylands. Challenge Paper.* The Global Drylands Initiative, UNDP Drylands Development Centre, Nairobi, Kenya.

Dobie, P. and Goumandakoye, M. 2005. *The Global Drylands Imperative: Achieving the Millennium Development Goals in the drylands of the world.* UNDP Drylands Development Centre, Nairobi, Kenya.

Dregne, H. E. and Chou, N. 1992. Global desertification and costs. In: Dregne, H.E. (ed.) *Degradation and restoration of arid lands*. Texas Technical University, Lubbock, Texas. pp. 249-282.

Dupire, M. 1962. *Peuls nomades. Etude descriptive des Wodaabe du Sahel Nigérien*. Institut d'Ethnologie, Paris, France.

Easterling, W.E., Aggarwal, P.K., Batima, P., Brander, K.M., Erda, L., Howden, S.M., Kirilenko, A., Morton, J., Soussana, J.-F., Schmidhuber, J. and Tubiello, F.N. 2007. Food, fibre and forest products. In: Parry, M.L., Canziani, O.F., Palutikof, J.P., van der Linden, P.J. and Hanson, C.E. (eds.) *Climate change 2007: impacts, adaptation and vulnerability. Contribution of Working Group II to the Fourth Assessment Report of the Intergovernmental Panel on Climate Change*. Cambridge University Press, Cambridge, UK. pp. 273-313. www.ipcc.ch/

ECA. 2008. *Sustainable development report on Africa: Five-year review of the implementation of the World Summit on Sustainable Development outcomes in Africa (WSSD+5)*. United Nations Economic Commission for Africa, Addis Ababa, Ethiopia.

Eckholm, E., Goley, G., Barnard, G. and Timberlake, L. 1984. *Firewood: the energy crisis that won't go away*. Earthscan, London, UK.

Ecklundh, L. and Olsson, L. 2003. Vegetation index trends for the African Sahel 1982-1999. *Geophysical Research Letters* **30**(8): 1430.

Eriksen, S., O'Brien, K., and Rosentrater, L. 2008. *Climate change in Eastern and Southern Africa: Impacts, vulnerability and adaptation*. GECHS Report 2008:2. University of Oslo, Oslo, Norway.

Eriksen, S.H., Brown, K. and Kelly, P.M. 2005. The dynamics of vulnerability: local coping strategies in Kenya and Tanzania. *The Geographical Journal* **17**(1/4): 287-305.

Eriksen, S., Ulsrud, K., Lind, J. and Muok, B. 2006. *The urgent need to increase adaptive capacities. Evidence from Kenyan drylands*. Conflicts and Adaptation Policy Brief 2. African Centre for Technology Studies, Nairobi, Kenya.

Faerge, J. and Magid, J. 2004. Evaluating NUTMON nutrient balancing in Sub-Saharan Africa. *Nutrient Cycling in Agro-ecosystems* **69**(2): 101-110.

Fan, S., Hazell, P. and Haque, T. 2000. Targeting public investments by agroecological zone to achieve growth and poverty alleviation goals in rural India. *Food Policy* **25**: 411-428.

FAO. 2000. *Global forest resources assessment. Main Report*. FAO Forestry Paper 140, Food and Agriculture Organisation of the United Nations, Rome, Italy.

FAO. 2005. Editorial: food security, complex emergencies and longer-term programming. *Disasters* **29**(1): 1-4.

FAO. 2008. *Are grasslands under threat? Brief analysis of FAO statistical data on grassland and fodder crops*. Food and Agriculture Organisation of the United Nations, Rome, Italy.

Faye, J. 2008. *Land and decentralisation in Senegal*. IIED Issues Paper 149. International Institute for Environment and Development, London, UK.

Faye, A., Fall, A., Mortimore, M., Tiffen, M. and Nelson, J. 2001. *Région de Diourbel: Synthesis*. Drylands Research Working Paper 23e. Drylands Research, Crewkerne, UK.

Foley, G. 2001. *Sustainable wood fuel supplies from the dry tropical woodlands*. ESMAP Technical Paper 013. The World Bank, Washington, DC.

Franke, R.W. and Chasin, B.H. 1980. *Seeds of famine: ecological destruction and the development dilemma in the West African Sahel*. Rowman/Allanheld, Totowa, New Jersey, USA.

Gabre-Madhin, E. and Haggblade, S. 2004. Success in African agriculture: results of an expert survey. *World Development* **32**(5):745-766.

Ghazi, P., Barrow, E., Monela, G. and Mlenge, W. 2005. Regenerating woodlands: Tanzania's HASHI Project. In: WRI *et al. The wealth of the poor: managing ecosystems to fight poverty*. World Resources Institute with UNDP, UNEP, The World Bank, Washington, DC. pp. 131-138.

Grace, J., San Jose, J., Meir, P., Miranda, H. and Montes, R. 2006. Productivity and carbon fluxes of tropical savannas. *Journal of Biogeography* **33**: 387-400.

GTZ. 2008. *A roadmap for biofuels in Kenya: Opportunities and obstacles*. Federal Ministry for Economic Cooperation and Development, Bonn, Germany.

Gunderson, L.H., Holling, C.S. and Light, S.S. 1995. *Barriers and bridges to the renewal of ecosystems and institutions*. Columbia University Press, New York, USA.

Haile, M., Witten, W., Abraha, K., Fissha, S., Kebede, A., Kassa, G. and Reda, G. 2005. *Land registration in Tigray, northern Ethiopia*. International Institute for Environment and Development, London, UK.

Harris, F.M.A. 1998. Farm-level assessment of the nutrient balance in northern Nigeria. *Agriculture, Ecosystems and Environment* **71**: 201-214.

Harris, F. and Mohammed, S. 2003. Relying on nature: wild foods in northern Nigeria. *Ambio* **32**(1): 24-29.

Harris, F. and Yusuf, M.A. 2001. Manure management by smallholder farmers in the Kano Close-Settled Zone, Nigeria. *Experimental Agriculture* **37**: 319-332.

Hatfield, R. and Davies, J. 2006, *Global review of the economics of pastoralism*. World Initiative for Sustainable Pastoralism/IUCN, Nairobi, Kenya.

Hazell, P. 2001. *Strategies for the sustainable development of dryland areas*. Global Drylands Partnership, International Food Policy Research Institute, Washington, DC.

Hazell, P., Jansen, H., Ruben, R. and Kuyvenhoven, A. 2002. *Investing in poor people in poor lands*. Paper prepared for IFAD, IFPRI, NIFP and Wageningen University and Research Centre, International Food Policy Research Institute, Washington, DC.

Helldén, U. and Tottrup, C. 2009. Regional desertification: a global synthesis. *Global and Planetary Change* **64**(3-4): 169-176.

Henao, J. and Banaante, C. 1999. *Estimating rates of nutrient depletion in soils of agricultural lands of Africa*. International Fertilizer Development Center, Muscle Shoals, Alabama, USA.

Herrmann, S.M., Anyamba, A. and Tucker, C.J. 2005. Recent trends in vegetation dynamics in the African Sahel and their relationship to climate. *Global Environmental Change* **15**: 394-404.

Hesse, C. and Thébaud, B. 2006. Will pastoral legislation disempower pastoralists in the Sahel? *Indigenous Affairs* no. **1**/06: 14-22.

Hill, P. 1972. *Rural Hausa: A village and a setting.* Cambridge University Press, Cambridge, UK.

Holling, C.S. 1973. Resilience and stability of ecological systems. *Annual Review of Ecology and Systematics* **4**: 1-23.

Holling, C.S. 2001. Understanding the complexity of economic, ecological, and social systems. *Ecosystems* **4**: 390-405.

Holmén, H. 2005. Spurts in production - Africa's limping green revolution. In: Djurfeldt, G. *et al.* (eds.) *The African food crisis. Lessons from the Asian green revolution.* CABI Publishing, Wallingford, UK. pp. 65-86.

Hyman, E. 1993. Forestry policies and programmes for fuelwood supply in northern Nigeria. *Land Use Policy* **10**(1): 26-43.

ICIMOD and WISP. 2007. *Report on experiences of ingenious rangeland co-management in northern Tibet, China.* International Centre for Integrated Mountain Development, Kathmandu, Nepal.

ICRAF. 2008. *World Agroforestry.* International Council for Research on Agroforestry, Nairobi, Kenya.

IFAD/FAO. 2008. *Water and the rural poor.* International Fund for Agricultural Development, Rome, Italy.

IISD. 2008. *A Summary Report of the Africa Carbon Forum.* International Institute for Sustainable Development: http://www.iisd.ca/africa/acf/

ILRI. 2005. *CGIAR Brief: Avian influenza and the developing world.* Unpublished report International Livestock Research Institute, Nairobi, Kenya.

ILRI. 2007. *Nature's benefits in Kenya. An atlas of ecosystems and human well-being.* World Resources Institute with ILRI, Washington, DC.

Ingram, K.K., Roncoli, M.C. and Kirshen, P.H. 2002. Opportunities and constraints for farmers of west Africa to use seasonal precipitation forecasts with Burkina Faso as a case study. *Agricultural Systems* **74**: 331-349.

IPCC. 2007. *Fourth Assessment of the Intergovernmental Panel on Climate Change: Summary for policy makers.* Intergovernmental Panel on Climate Change. www.ipcc.ch/

IRI. 2005. *Sustainable development in Africa. Is the climate right?* International Research Institute for Climate Prediction, Columbia University, New York, USA.

IUCN. 2007. *The business of Mongongo in Zambia: A case study from the Natural Futures Programme.* IUCN, Pretoria, South Africa.

Jaspars, S. 2009. *From food crisis to fair trade: Livelihoods analysis, protection and support in emergencies.* Oxfam/ Emergency Nutrition Network, Oxford, UK.

Jouet, A., Jouve, P., and Banoin, M. 1996, Le défrichement amélioré au Sahel. Une pratique agroforestiére adoptée par les paysans. In: Jouve, P. (ed.) *Gestion des terroirs et des ressources naturelles au Sahel.* CNEARC, Montpellier, France. pp. 34-42.

Knowler, D., Acharya, G. and van Rensburg, T. 1998. *Incentives systems for natural resources management.* Investment Centre, Food and Agriculture Organisation, Rome, Italy.

Krätli, S. 2008. *Time to outbreed animal science? A cattle breeding system exploiting structural unpredictability. The WoDaaBe herders in Niger.* STEPS Working Paper 7. Institute of Development Studies, Brighton, UK..

Lal, R. 2000. Soil management in the developing countries. *Soil Science* **165**(1): 57-72.

Lane, C. 1998. *Custodians of the commons. Pastoral land tenure in East and West Africa.* Earthscan Publishers, London, UK.

Laris, P. and Wardell, D.A. 2006. Good, bad or 'necessary evil'? Reinterpreting the colonial burning experiments in the savanna landscapes of West Africa. *The Geographical Journal* **172**(4): 271-290.

Lemenih, M., Abebe, T. and Olsson, M. 2003. Gums and resin resources from some *Acacia, Boswellia* and *Commiphera* species and their economic contributions in Liban, south-east Ethiopia. *Journal of Arid Environments* **55**(465): 482.

Lemenih, M., Feleke, S. and Tadesse, W. 2007. Constraints to smallholders production of frankincense in Metema district, north-west Ethiopia. *Journal of Arid Environments* **71**: 393-403.

Lepers, E., Lambin, E.F., Janetos, A., DeFries, R., Chard, F., Ramankutty, N. and Scholes, R.J. 2005. A synthesis of information on rapid land-cover change for the period 1981-2000. *Bioscience* **55**(2): 115-124.

Linares-Palomino, R. 2009. *Synthesis of status of drylands ecosystem services in South and Central America.* Unpublished report to IUCN, Nairobi, Kenya.

Liu, J., Dietz, T., Carpenter, S.R., Alberti, M., Folke, C., Moran, E., Pell, A.N., Deadman, P., Kratz, T., Lubchenko, J., Ostrom, E., Ouyang, Z., Provencher, W., Redman, C.L., Schneider, S.H. and Taylor, W.W. 2007. Complexity of coupled human and natural systems. *Science* **317**: 1513-1516.

Lo, H. and Dione, M. 2000. *Région de Diourbel: Evolution des régimes fonciers.* Drylands Research Working Paper 19. Drylands Research, Crewkerne, UK.

Maass, J. *et al.*, 2005. Ecosystem services of tropical dry forests: insights from long-term ecological and social research on the pacific coast of Mexico. *Ecology and Society* **10**(1): 17.

Madzwamuse, M., Schuster, B., Nherera, B., and Kerven, C. 2007. *The real jewels of the Kalahari. Dryland goods and services in Kgalagadi South District, Botswana.* IUCN, Gland, Switzerland.

Mahamane, A. 2001. *Usages des terres et évolutions végétales dans le département de Maradi.* Drylands Research Working Paper 27. Drylands Research, Crewkerne, UK.

Mahul, O. and Skees, J. 2006. Piloting Index-Based Livestock Insurance in Mongolia. *Access Finance* **10**: 1-4. A newsletter published by the financial sector vice presidency. The World Bank Group, Washington, DC.

Mando, A., Fatondji, D., Zougmoré, R., Brusaard, L., Bielders, C. L. and Martius, C. 2006. Restoring soil fertility in semi-arid West Africa: assessment of an indigenous technology. In: Uphoff, N. (ed.) *Biological approaches to sustainable soil systems*. CRC Taylor and Francis, New York, USA. pp. 391-400.

Maranz, S. 2009. Tree mortality in the African Sahel indicates an anthropogenic system displaced by climate change. *Journal of Biogeography* **36**: 1181-1193.

Marshall, E., Schreckenberg, K. and Newton, A. C. (eds.) 2006. *Commercialization of non-timber forest products: factors influencing success. Lessons learnt from Mexico and Bolivia and policy implications for decision makers*. UNEP World Conservation Monitoring Centre, Cambridge, UK.

Mazzucato, V. and Niemeijer, D. 2000. *Rethinking soil and water conservation in a changing society*. Tropical resource management paper 32. Wageningen University and Research Centre, Wageningen, The Netherlands.

McAuslan, P. 2006. *Improving tenure security for the poor in Africa. Framework paper for the legal empowerment workshop - sub-Saharan Africa*. Legal Empowerment of the Poor Working Paper #1. Food and Agriculture Organization of the United Nations, Rome, Italy.

MEA. 2005. *The Millennium Ecosystem Assessment*. World Resources Institute, Washington, DC. www:millenniumassessment.org/

Meagher, K. 1997. *Current trends in cross-border grain trade between Nigeria and Niger*. IRAM, Paris, France.

Menon, S. (ed.) 2008. *Managing drylands: Issues and perspectives*. Icfai University Books, Punjagutta, Hyderabad, India.

Meynen, W. and Doornbos, M. 2005. *Decentralising natural resource management: a recipe for sustainability and equity?* In: Ribot, J.C. and Larson, A. (eds.) *Decentralisation through a natural resource lens*. Routledge, London, UK. pp. 235-254.

Meles, K., Nigussie, G., Belay, T. and Manjur, K. 2009. *Seed system impact on farmers' income and crop biodiversity in the drylands of southern Tigray*. DCG Report no. 54, Drylands Coordination Group, Oslo, Norway.

Meze-Hausken, E. 2000. Migration caused by climate change: how vulnerable are people in dryland areas? *Mitigation and Adaptation Strategies for Global Change* **5**: 379-406.

Meze-Hausken, E. 2004. Contrasting climate variability and meteorological drought with perceived drought and climate change in northern Ethiopia. *Climate Research* **27**: 19-31.

Mortimore, M. 1989. *Adapting to drought, farmers, famines and desertification in West Africa*. Cambridge University Press, Cambridge, UK.

Mortimore, M. 1993a. Northern Nigeria: land transformation under agricultural intensification. In: Jolly, C.L, and Torrey, B.B. (eds.) *Population and land use in developing countries. Report of a workshop*. National Academic Press, Washington, DC. pp. 42-69.

Mortimore, M. 1993b. The intensification of peri-urban agriculture: the Kano Close-Settled Zone, 1964-86. In: Turner II, B.L, Kates, R.W. and Hyden, H.L. (eds.) *Population growth and agricultural change in Africa*. University Press of Florida Gainesville. pp. 358-400.

Mortimore, M. 1998. *Roots in the African dust: sustaining the Sub-Saharan drylands*. Cambridge University Press, Cambridge, UK.

Mortimore, M. 2003. *The future of family farms in West Africa: what can we learn from long-term data?* Drylands Research for the International Institute of Environment and Development, London, UK.

Mortimore, M. 2005. *Why invest in drylands?* Global Mechanism of the UNCCD, Rome, Italy.

Mortimore, M. and Adams, W. 1999. *Working the Sahel: Environment and society in Northern Nigeria*. Routledge, London, UK.

Mortimore, M. and Harris, F. 2005. Do small farmers' achievements contradict the nutrient depletion scenarios for Africa? *Land Use Policy* **22**: 43-56.

Mortimore, M. and Tiffen, M. 1994. Population growth and a sustainable environment: the Machakos story. *Environment* **36**(8): 10-32.

Mortimore, M., Ariyo, J., Bouzou, I. B., Mohammed, S. and Yamba, B. 2008. Niger and Nigeria: the Maradi-Kano region. A case study of local natural resource management. In: Shepherd, G. (ed.) *The ecosystem approach. Learning from experience*. IUCN, Gland, Switzerland. pp. 44-58.

Mortimore, M., Ba, M., Mahamane, A., Rostom, R.S., Serra del Pozo, P. and Turner, B. 2005. Changing systems and changing landscape: measuring and interpreting land use transformations in African drylands. *Geografisk Tidsskrift. Danish Journal of Geography* **105**(1): 101-118.

Mortimore, M., Harris, F. and Turner, B. 1999. Implications of land use change for the production of plant biomass in densely populated Sahelo-Sudanian shrub-grasslands in north-east Nigeria. *Global Ecology and Biogeography* **8**: 243-256.

Morton, J. 2006. Pastoralist coping strategies and emergency livestock interventions. In: McPeak, J.G. and Little, P.D. (eds.) *Livestock marketing in Eastern Africa: research and policy challenges*. ITDG Publications, London, UK.

Morton, J. 2007. The impact of climate change on smallholder and subsistence agriculture. *Proceedings of the National Academy of Sciences* **104**: 19680-19685.

Munasinghe, M. 2009. *Sustainable development in practice: Sustainomics methodology and applications*. Cambridge University Press, Cambridge, UK.

Musemwa, L.A., Mushunje, A., Chimonyo, M., Fraser, G., Mapiye, C. and Muchenje, V. 2008. Nguni cattle marketing constraints and opportunities in the communal areas of South Africa: a review. *African Journal of Agricultural Research* **3**(4): 239-245.

Mustapha, A. R. and Meagher, K. 2000. *Agrarian production, public policy and the State in Kano Region*, 1900-2000. Drylands Research Working Paper 35. Drylands Research, Crewkerne, UK.

Muyakwa, S. L. 2009. *Study report on the impact of Jatropha and cotton farming on the livelihoods of farmers and the environment in Zambia.* Organisation Development and Community Management Trust, Lusaka, Zambia.

Ngoitiko, M. 2008. *The Pastoral Women's Council: Empowerment for Tanzania's Maasai.* Gatekeeper 137e. International Institute for Environment and Development, London, UK.

Niamir-Fuller, M. (ed.) 1999. *Managing mobility in African rangelands: The legitimisation of transhumance.* Intermediate Technology Publications, London, UK.

Nori, M. 2004, *Hoofs on the roof: Pastoral livelihoods on the Qinghai-Tibetan plateau. The case of Chengduo county, Yushu prefecture.* Association for International Solidarity in Asia (ASIA).

Nori, M., Taylor, M. and Sensi, A. 2008. *Browsing on fences. Pastoral land rights, livelihoods and adaptation to climate change.* IIED Issue Paper 148. International Institute for Environment and Development, London, UK.

Norton-Griffiths, M. 2007. How many wildebeest do you need? *World Economics* **8**(2): 41-64.

Nyoni, D. 2008. *The Organisation of Rural Associations for Progress, Zimbabwe: Self-reliance for Sustainability.* Gatekeeper 137d. International Institute for Environment and Development, London, UK.

Ojima, D. and Chuluun, T. 2008. Policy changes in Mongolia: implications for land use and landscapes. In: Galvin, K. A., Reid, R. S., Behnke, R. H., and Hobbs, N. T. (eds.) *Fragmentation in semi-arid and arid landscapes: consequences for human and natural systems.* Springer, Dordrecht, Germany. p. 179-193.

Oldeman, R. and Hakkeling, R. 1990. *World map of the status of human-induced soil degradation: an explanatory note.* United Nations Environment Programme, Nairobi, Kenya.

Olsson, L., Eklundh, L. and Ardö, J. 2005. A recent greening of the Sahel - trends, patterns and potential causes. *Journal of Arid Environments* **63**(3): 556-566.

PAGRNAT. 2001. *Aide mémoire de la mission conjointe de formulation, 5-10 février 2001.* Danida, UNDP-DDC, IUCN, DFPP, Niamey, Niger.

PAGRNAT. 2002. *Identification d'un nouveau programme en faveur des populations pastorales de l'Aïr -Ténéré, 4-17 février 2002.* Mission d'appui de la Coopĕration Suisse, Niamey, Niger.

Peru. 2008. *Informe Perú a la XXVII Reunión Ordinaria del Convenio para la Conservación de la Vicuca. Decreto Ley No.22984.* CONACS, Lima, Peru.

Rahmato, D. 1991. *Famine survival strategies: A case study from northeast Ethiopia.* Scandinavian Institute of African Studies, Uppsala, Sweden.

Rapport, D. 2009. Healthy ecosystems: an evolving paradigm. In: Pretty, J. *et al.* (eds.) *Sage handbook on environment and society.* Sage, London, UK.

Raynaut, C. 1980. *Recherches multidisciplinaires sur la région de Maradi: Rapport de Synthuse.* Programme de Recherche de la Région de Maradi, Université de Bordeaux II, Bordeaux, France.

Reardon, T. and Vosti, S.A. 1992. Issues in the analysis of the effects of policy on conservation and productivity at the household level in developing countries. *Quarterly Journal of International Agriculture* **31**(4): 380-396.

Reardon, T., Timmer, C.P., Barrett, C. and Berdegue, J. 2003. The rise of supermarkets in Africa, Asia and Latin America. *American Journal of Agricultural Economics* **85**(5): 1140-1146.

Reenberg, A., Nielsen, T.L. and Rasmussen, K. 1998. Field expansion and reallocation in the Sahel - land use pattern dynamics in a fluctuating biophysical and socio-economic environment. *Global Environmental Change* **4**: 309-327.

Reij, C. and Steeds, D. 2003. *Success stories in Africa's drylands: supporting advocates and answering sceptics.* CIS/Centre for International Cooperation, Vrije Universiteit, Amsterdam, The Netherlands.

Reij, C. and Thiombiano, T. 2003. *Développement rural et environnement au Burkina Faso: la réhabilitation de la capacité productive des terroirs sur la partie nord du Plateau Central entre 1980 et 2001.* CIS, Vrije Universiteit, Amsterdam, The Netherlands.

Reynolds, J.F., Stafford Smith, D.M., Lambin, E.F., Turner, B.L.I., Mortimore, M., Batterbury, S.P.J., Downing, T.E., Dowlatabadi, H., Fernández, R.J., Herrick, J.E., Huber-Sannwald, E., Jiang, H., Leemans, R., Lynam, T., Maestre, F.T., Ayarza, M. and Walker, B. 2007. Global desertification: Building a science for dryland development. *Science* **316**: 847-851.

Ribot, J.C. 1995. From exclusion to participation: turning Senegal's Forestry Policy around. *World Development* **23**(9): 1587-1599.

Roncoli, C., Ingram, K. and Kirshen, P. 2001. The costs and risks of coping with drought: livelihood impacts and farmers' responses in Burkina Faso. *Climate Research* **19**: 119-132

Rosenzweig, C., Casassa, G., Karoly, D. J., Imeson, A., Liu, C., Menzel, A., Rawlins, S., Root, T. L., Seguin, B. and Tryjanovski, P. 2007. Assessment of observed changes and responses in natural and managed ecosystems. In: Parry, M.L., Canziani, O.F., Palutikof, J.P., van der Linden, P.J. and Hanson, C.E. (eds.) *Climate change 2007. Impacts, adaptation, vulnerability. Contribution of Working Group II to the Fourth Assessment of the Intergovernmental Panel on Climate Change*, Cambridge University Press, Cambridge, UK. pp. 79-131. www.ipcc.ch/

Sadik, N. 1991. Population growth and the food crisis. *Food, nutrition and agriculture* **1**(1): 3-6.

Safriel, U., Adeel, Z., Niemeijer, D., Puigdefabregas, J., White, R., Lal, R., Winslow, M., Ziedler, J., Prince, S., Archer, E., and King, C. 2005. Chapter 22: Dryland systems. In: Hassan, R., Scholes, R. and Ash, N. (eds.) *Millennium Ecosystem Assessment. Vol. 1. Ecosystems and human well-being: Current state and trends.* World Resources Institute, Washington, DC. pp. 623-662.

Sandford, S. 1983. *Management of pastoral development in the Third World.* John Wiley. Chichester, UK.

Sandford, S. 1994. Improving the efficiency of opportunism: new directions for pastoral development. In: Scoones, I. (ed.) *Living with uncertainty. New directions in pastoral development.* Intermediate Technology Publications, London, UK. pp. 174-182.

Sankaran, M. *et al.* 2005. Determinants of woody cover in African savannas. Letter in *Nature* **438**: 846-849.

Schlecht, E., Fernandez-Rivera, S. and Hiernaux, P. 1998. Timing, size and N-concentration of faecal and urinary excretions in cattle, sheep and goats: Can they be exploited for better manuring of cropland? In: Renard, G., Neef, A., Becker, K. and Von Oppen, M. (eds.) *Soil fertility management in West African land use systems.* Margraf Verlag, Weikersheim, Germany. pp. 361-368.

Scoones, I. and Toulmin, C. 1998. Soil nutrient balances: what use for policy? *Agriculture, Ecosystems and Environment* **71**: 255-267.

Scoones, I. (ed.) 1994. *Living with uncertainty. New directions in pastoral development in Africa.* Intermediate Technology Publications, London, UK.

Shanahan, T.M., Overpeck, J.T., Anchukaitis, K.J., Beck, J.W., Cole, J.E., Dettman, D.L., Pech, J.A., Scholz, C.A. and King, J.W. 2009. Atlantic forcing of persistent drought in West Africa. *Science* **324**(5925): 377-380

Solomon, S., Qin, D., Manning, M., Chen, Z., Marquis, M., Averyt, K. B., Tignor, M. and Miller, H. (eds.) 2007. *Climate change 2007. The physical science basis. Contribution of Working Group I to the Fourth Assessment Report of the Intergovernmental Panel on Climate Change.* Cambridge University Press, Cambridge, UK.

Stebbing, E.P. 1953. *The creeping desert in the Sudan and elsewhere in Africa 15 to 30 degrees latitude.* McCorqudale, Khartoum, Sudan.

Stenning, D.J. 1959. *Savannah nomads. A study of the Wodaabe pastoral Fulani of western Bornu Province Northern Region Nigeria.* Oxford University Press, Oxford, UK.

Stern, N. 2007. *Stern Review on the economics of climate change. Executive Summary.* HM Treasury, London, UK.

Stoorvogel, J. J. and Smaling, E. M. A. 1990. *Assessment of soil nutrient depletion in Sub-saharan Africa: 1983-2000. Vol.1: Main Report (second edition).* Report 28, Winand Staring Centre, Wageningen, The Netherlands.

Sullivan, C. A. and O'Regan, D. P. 2003. *Winners and losers in forest product commercialisation. Final Report. Vol 1.* Centre for Ecology and Hydrology, Wallingford, UK and the Department for International Development, London, UK.

t'Mannetjie, L., Amézquita, M.C., Buurman, P., and Ibrahim, M.A. (eds.) 2008. *Carbon sequestration in tropical grassland ecosystem.* Wageningen Academic Publishers, Wageningen, The Netherlands.

Tadesse, W., Desalegn, G. and Alia, R. 2007. Natural gum and resin bearing species of Ethiopia and their potential applications. *Investigación Agraria: Sistemas y Recursos Forestales* **16**(3): 211-221.

Tennigkeit, T. and Wilkes, A. 2008. *An assessment of the potential for carbon finance in rangelands. ICRAF Working Paper 68.* World Agroforestry Centre, Southeast Asia.

Thébaud, B. 2002. *Foncier pastoral et gestion de l'espace au Sahel: Peuls du Niger oriental et du Yagha burkinabé.* Karthala, Paris, France.

Thébaud, B., Grell, H. and Miehe, S. 1995. *Recognising the effectiveness of traditional pastoral practices: lessons from a controlled grazing experiment in northern Senegal.* IIED Issue Paper 55. International Institute for Environment and Development, London, UK.

Thomas, R.J. 2008. Opportunities to reduce the vulnerability of dryland farmers in Central and West Asia and North Africa to climate change. *Agriculture, Ecosystems and Environment* **126**: 36-45.

Thornton, P. K., Jones, P. G., Owiyo, T., Kruska, R. L., Herrero, M., Kristjanson, P., Notenbaert, A., Bekele, N. and Omolo, A. 2006. *Mapping climate vulnerability and poverty in Africa.* Report to the Department for International Development, ILRI, Nairobi, Kenya.

Tiffen, M. and Mortimore, M. 2002. Questioning desertification in dryland Africa sub-Saharan Africa. *Natural Resources Forum* **26**(3): 218-233.

Tiffen, M., Mortimore, M. and Gichuki, F. 1994. *More people less erosion: Environmental recovery in Kenya.* John Wiley & Sons, Chichester, UK.

Toulmin, C. 1992. *Cattle, women and wells. Managing household survival in the Sahel.* Clarendon Press, Oxford, UK.

Toulmin, C. and Guèye, B. 2003. *Transformation in West African agriculture and the role of family farms.* Issue Paper 123. International Institute for Environment and Development, London, UK.

Toulmin, C. and Quan, J. (eds.) 2000. *Evolving land rights, policy and tenure in Africa.* International Institute for Environment and Development, London, UK.

Trenberth, K. E., Jones, P. D., Ambenje, P., Bojanu, R., Easerling, D., Klein Tank, A., Parker, D., Rahimzadeh, F., Renwick, J. A., Rusticucci, M., Soden, B. and Zhai, P. 2007. Observations: surface and atmospheric climate change. In: Solomon, S., Qin, D., Manning, M., Chen, Z., Marquis, M., Averyt, K. B., Tignor, M. and Miller, H. (eds.) *Climate change 2007. The physical science basis. Contribution of Working Group I to the Fourth Assessment Report of the Intergovernmental Panel on Climate Change.* Cambridge University Press, Cambridge, UK. pp. 235-336.

Tschakert, P. 2007. Views from the vulnerable: Understanding climatic and other stressors in the Sahel. *Global Environmental Change* **17**(34): 381-396.

Tucker, C.J., Dregne, H.E. and Newcomb, W.W. 1991. Expansion and contraction of the Sahara Desert from 1980 to 1990. *Science* **253**(5017): 299-301.

Turner, M.D. 1999a. Merging local and regional analyses of land-use change: the case of livestock in the Sahel. *Annals of the Association of American Geographers* **89**: 191-219.

Turner, M.D. 1999b. The role of social networks, indefinite boundaries and political bargaining in maintaining the ecological and economic resilience of the transhumance systems of Sudano-Sahelian West Africa. In: Niamir-Fuller, M. (ed.) *Managing mobility in African rangelands: The legitimisation of transhumance.* Intermediate Technology Publications, London, UK.

UNCCD. 1993. *United Nations Convention on Combating Desertification.* United Nations Convention to Combat Desertification, Bonn, Germany.

UNCCD. 2006. *Implementing the United Nations Convention to Combat Desertification in Africa. Ten African experiences.* UNCCD, Bonn, Germany.

Undargaa, S. 2006. *Gender and pastoral land use in Mongolia: Dilemmas of pastoral land tenure.* The Centre for Development Studies. The University of Auckland, Auckland, New Zealand.

UNDP. 2004. *Reducing disaster risk: A challenge for development.* United Nations Development Programme, New York, USA.

UNDP-DDC. 2001. *Pastoralism and mobility in the drylands. Challenge Paper.* UNDP Drylands Development Centre, Nairobi, Kenya.

UNDP-DDC. 2003. *Land Tenure Reform and the Drylands. Challenge Paper.* The Global Drylands Initiative, UNDP Drylands Development Centre, Nairobi, Kenya.

UNDP-DDC. 2006. *2nd African Drought Risk and Development Forum Report, 16-18 October 2006.* UNDP Drylands Development Centre, Nairobi, Kenya.

UNDP-DDC. 2008. *3rd African Drought Adaptation Forum Report. Addis Ababa, Ethiopia, 17-19 September 2008.* UNDP Drylands Development Centre, Nairobi, Kenya.

UNEP. 1977. *Report of the United Nations Conference on Desertification, 29 August - 9 September 1977.* United Nations Environment Programme, Nairobi, Kenya.

UNEP. 1992. *World Atlas of Desertification.* 1st ed. Edward Arnold, London, UK.

UNEP. 2007. *Sudan post-conflict environmental assessment.* United Nations Environment Programme, Nairobi, Kenya.

UNEP/GEF. 2005. *Integrated ecosystem management of transboundary areas between Niger and Nigeria phase I: Strengthening of legal and institutional frameworks for collaboration and experimental demonstration of integrated ecosystem management.* United Nations Environment Programme/Global Environmental Facility project GEF/GFL 2328-2770-4889.

Uphoff, N., Ball, A.S., Fernandes, E., Herren, H., Husson, O., Laing, M., Palm, C., Pretty, J., Sanchez, P., Sanginga, N. and Thies, J. 2006. *Biological approaches to sustainable soil systems.* CRC Press, Boca Raton, Florida, USA.

Van den Eynden, V., Cueva, E. and Cabrero, O. 2003. Wild foods from southern Ecuador. *Economic Botany* **57**(4): 576-603.

Van Duivenbooden, N., Abdoussallam, S. and Ben Mohammed, A. 2002. Impact of climate change on agricultural production in the Sahel. Part 2: case study for groundnut and cowpea in Niger. *Climate Change* **54**(349): 368.

Vermeulen, S., Woodhill, J., Proctor, F.J. and Delnoye, R. 2008. *Chain-wide learning for inclusive agrifood market development: a guide to multi-stakeholder processes for linking small-scale producers with modern markets.* International Institute for Environment and Development, London, UK and Wageningen University and Research Centre, Wageningen, The Netherlands.

Vogt, G. and Vogt, K. 2000. *Hannu biu ke tchudu juna - strength in unity: a case study from Takieta, Niger.* Securing the Commons Series no. 2. IIED/SOS Sahel, London, UK.

Vosti, S. and Reardon, T. (eds.) 1997. *Sustainability, growth and poverty alleviation: a policy and agroecological perspective.* International Food Policy Research Institute. John Hopkins University Press, Baltimore, Maryland, USA.

Wane, A., Toutain, B., Touré, I., Diop, A. T. and Ancey, V. 2008. *Pastoralism as an economic mode of valorization of the arid areas. A case study at the site of Tatki (in the Senegalese Sahel).* Working Paper in press, PPZS, Dakar, Senegal.

Watts, M.J. 1983. *Silent violence: food, famine and the peasantry in Northern Nigeria.* University of California, Berkeley, USA.

Western, D. 1982. The environment and ecology of pastoralism in arid savannas. *Development and Change* **13**: 183-211.

Wilkes, A. 2008. *Towards mainstreaming climate change in grassland management policies and practices on the Tibetan Plateau.* ICRAF Working Paper no. 68. World Agroforestry Centre, Southeast Asia.

Wily, L.A. 2006. *Land rights reform and governance in Africa. How to make it work in the 21st century?* UNDP Drylands Development Centre, Nairobi, Kenya and UNDP Oslo Governance Centre, Oslo, Norway.

WISP. 2007a. *Pastoralism as conservation in the Horn of Africa.* World Initiative for Sustainable Pastoralism, Nairobi, Kenya.

WISP. 2007b. *Pastoral institutions for managing natural resources and landscapes. WISP Policy Brief No 6.* World Initiative for Sustainable Pastoralism, Nairobi, Kenya.

World Bank. 2000. *World Development Report 2000/2001: Attacking poverty.* The World Bank, Washington, DC.

World Bank. 2003. *Sustainable development in a dynamic world. Transforming institutions, growth, and quality of life. World Development Report 2003.* The World Bank, Washington, DC.

World Bank. 2007. *Agriculture for development. World Development Report 2008.* The World Bank, Washington, DC.

World Bank. 2008. *Lessons from World Bank Group responses to past financial crises. Independent Evaluation Group Evaluation Brief 6*. The World Bank, Washington, DC.

WRI *et al.* 2005. *The wealth of the poor. managing ecosystems to fight poverty. WRI Annual Report*. World Resources Institute with UNDP, UNEP, The World Bank, Washington, DC.

WRI. 2008. *World Resources 2008: Roots of resilience – Growing the wealth of the poor*. World Resources Institute, Washington, DC.

WRI *et al.* 2007. *Nature's benefits in Kenya. An atlas of ecosystems and human well-being*. World Resources Institute, Washington, DC and Nairobi, Kenya.

Zonon, A., Kerven, C. and Behnke, R. 2007. *A preliminary assessment of the economic value of the goods and services provided by dryland ecosystems of the Anr and* Ténéré. IUCN, Gland, Switzerland.

Photo Credits

Cover: IUCN Photo Library/ Danièle Perrot-Maître; p. 3 Chris Reij; p. 9 Jal Bhagirathi Foundation; p. 16 Caterina Wolfangel; p. 16 Edmund Barrow; p. 19 Piet Wit; p. 26 Christopher Taylor; p. 32 Caterina Wolfangel; p. 34 Royer / Still Pictures; p. 38/39 Wendy Strahm; p. 45 Agni Boedhihartono; p. 50 Daniel Kreuzberg; p. 54 Agni Boedhihartono; p. 60 IUCN Photo Library/ Michael Mortimore; p. 65 IUCN Photo Library/ Michael Mortimore; p. 72 Jonathan Davies.